U.S. Department
of Transportation
National Highway
Traffic Safety
Administration

DOT HS 810 794
NHTSA Technical Report

July 2007

Statistical Analysis of the Effectiveness of Electronic Stability Control (ESC) Systems – Final Report

---

1. Report No. DOT HS 810 794	2. Government Accession No.	3. Recipient's Catalog No.
4. Title and Subtitle STATISTICAL ANALYSIS OF THE EFFECTIVENESS OF ELECTRONIC STABILITY CONTROL (ESC) SYSTEMS – FINAL REPORT		5. Report Date July 2007
		6. Performing Organization Code
7. Author(s) Jennifer N. Dang		8. Performing Organization Report No.
9. Performing Organization Name and Address Evaluation Division; National Center for Statistics and Analysis National Highway Traffic Safety Administration Washington, DC 20590		10. Work Unit No. (TRAIS)
		11. Contract or Grant No.
12. Sponsoring Agency Name and Address Department of Transportation National Highway Traffic Safety Administration Washington, DC 20590		13. Type of Report and Period Covered NHTSA Technical Report
		14. Sponsoring Agency Code
15. Supplementary Notes		

16. Abstract

Electronic Stability Control (ESC) is a safety technology designed to enhance a vehicle's stability and control in all driving situations ESC first became available in the United States in 1997 Statistical analyses of 1997-2004 crash data from the Fatality Analysis Reporting System (FARS) and 1997-2003 crash data from the State data files estimate reductions with ESC for various types of crash involvements

- ESC reduced **fatal run-off-road** crashes by 36 percent for passenger cars and 70 percent for light trucks and vans (LTVs) The reductions are statistically significant
- **Police-reported run-off-road** involvements were decreased by 45 percent in passenger cars and 72 percent in LTVs The decreases are statistically significant
- **Fatal single-vehicle** crashes that did not involve pedestrians, bicycles, and animals decreased (due to ESC) by 36 percent in passenger cars and 63 percent in LTVs The decreases are statistically significant
- ESC reduced **police-reported single-vehicle** crashes (excluding pedestrian, bicycle, animal crashes) by 26 percent for passenger cars and 48 percent for LTVs The reductions are statistically significant
- **Rollover** involvements in fatal crashes were decreased by 70 percent in passenger cars and 88 percent in LTVs The decreases are statistically significant
- **Police-reported crashes involving rollovers** were reduced by 64 percent in passenger cars and 85 percent in LTVs The reductions are statistically significant
- ESC reduced **culpable fatal multi-vehicle** crashes by 19 percent for passenger cars and 34 percent for LTVs Only the reduction involving LTVs is statistically significant
- **Culpable involvements in police-reported multi-vehicle** crashes were decreased by 13 percent in passenger cars and 16 percent in LTVs The decreases are statistically significant
- **Overall**, ESC reduced all fatal crashes by 14 percent for passenger cars and 28 percent for LTVs Only the reduction in LTVs is statistically significant
- **Overall, police-reported crash involvements** decreased by 8 percent in passenger cars and 10 percent in LTVs The decreases are statistically significant

17. Key Words NHTSA; FARS; ESC; statistical analysis; evaluation; benefits; effectiveness; fatality reduction; electronic stability control systems;	18. Distribution Statement Document is available to the public at the Docket Management System of the U.S. Department of Transportation, http://dms.dot.gov, Docket Number 28629.

19. Security Classif. (Of this report) Unclassified	20. Security Classif. (Of this page) Unclassified	21. No. of Pages 63	22. Price

Form DOT F 1700.7 (8-72)   **Reproduction of completed page authorized**

TABLE OF CONTENTS

LIST OF ABBREVIATIONS ……………………………………………………iv

ACKNOWLEDGEMENTS …………………………………………………….v

EXECUTIVE SUMMARY ……………………………………………....… vi

BACKGROUND ……………………………………………………......1

ESC INFORMATION ……………………………………………....…5

CRASH DATA (STATE AND FARS) …………………………………… 11

ANALYSIS DATABASES ………………………………………………… 12

VIN decode …………………………………………………………..12
Crash involvements ……………………………………………........… 13

2X2 CONTINGENCY TABLE ANALYSIS …………………………………...18

Analysis of fatal crashes (FARS) …………………………………........... 20
Analysis of crash involvements (State Data) …………………………….25

LOGISTIC REGRESSION ANALYSIS ………………………………… 46

COMPARISON OF 2-CHANNEL AND 4-CHANNEL ESC SYSTEMS……………48

CONCLUSIONS……………………………………………………… 52

LIST OF ABBREVIATIONS

ABS	Antilock brake systems
ASE	Asymptotic standard error
CATMOD	Categorical models procedure in SAS
ESC	Electronic Stability Control
FARS	Fatality Analysis Reporting System
FMVSS	Federal Motor Vehicle Safety Standard
GENMOD	Generalized estimating models procedure in SAS
GM	General Motors
IIHS	Insurance Institute for Highway Safety
LR	Likelihood-ratio
LTVs	Light trucks and vans
MEANS	MEANS (simple statistics) procedure in SAS
NCSA	National Center for Statistics and Analysis
NHTSA	National Highway Traffic Safety Administration
OMB	Office of Management and Budget
SAS	Statistical analysis software produced by SAS Institute, Inc.
SE	Standard error
SUV	Sport utility vehicle
TCS	Traction control systems
VIN	Vehicle Identification Number

ACKNOWLEDGEMENTS

I owe special thanks to the two researchers who peer-reviewed a draft of the report:

1) Dr. Charles M. Farmer, Director of Statistical Services, Insurance Institute for Highway Safety, Arlington, Virginia

2) Professor Claes Tingvall, Dr Med Sc, Director of Traffic Safety at Swedish Road Administration

This study estimates the effectiveness of Electronic Stability Control (ESC) in reducing crashes, specifically crashes where ESC is likely to have made a difference in the vehicle's involvement, based on statistical analyses of crash data. The National Highway Traffic Safety Administration (NHTSA) published a draft of this report in support of a proposed rulemaking to establish a new Federal Motor Vehicle Safety Standard, FMVSS No. 126, which requires ESC systems on passenger cars, multipurpose vehicles, trucks, and buses with a gross vehicle weight rating of 10,000 pounds or less (See NHTSA Docket Number 25801-02 at http://dms.dot.gov). Because of the potential impacts of the proposed regulation, the report contains "highly influential scientific information" as defined by the Office of Management and Budget's (OMB) "Final Information Quality Bulletin for Peer Review" (available at www.whitehouse.gov/omb/inforeg/peer2004/peerbulletin.pdf). Therefore, the report had to be peer-reviewed in accordance with the requirements of both Sections II and III of OMB's Bulletin.

The peer-review process differed from the type used by journals. The effort by Dr. Farmer and Professor Tingvall was essentially assessment of the scientific adequacy of the draft to identify weaknesses and help NHTSA strengthen the report. These two reviewers were selected by NHTSA staff. The publication of this report does not necessarily imply that the reviewers supported it or concurred with the findings. You may access their comments on the draft along with the revised report and the entire review process in the NHTSA docket (Number 26415) at http://dms.dot.gov. We have tried to address all of the comments in our revised report (but we did not send it back to the reviewers for a second round of review). The text and footnotes of the report single out some of the reviewers' comments that instigated additions or revisions to the analyses.

EXECUTIVE SUMMARY

Electronic Stability Control (ESC) is a safety system designed to recognize adverse driving conditions by 1) continuously measuring and evaluating the speed, the steering wheel angle, the yaw rate, and the lateral acceleration of a vehicle from various sensors and 2) using those measured data to compare a driver's steering input with the vehicle's actual motion. If an unstable situation is detected, then ESC automatically intervenes to assist the driver and stabilize the vehicle by applying the brakes to individual wheels as needed and possibly reducing engine torque. This technology is expected to reduce the number of crashes due to driver error and loss of control, because it has the potential to anticipate situations leading up to some crashes before they occur and the capability to automatically intervene to prevent them. A major benefit should be the reduction of single-vehicle crashes that involve losing control and running off the road.

In September, 2004, the National Highway Traffic Safety Administration (NHTSA) issued an evaluation note on the *Preliminary Results Analyzing the Effectiveness of Electronic Stability Control (ESC) Systems*. The data suggested that ESC was highly effective in reducing single-vehicle run-off-road crashes. The study was based on Fatality Analysis Reporting System (FARS) data from calendar years 1997-2003 and crash data from five States from calendar years 1997-2002. The data were limited to mostly luxury vehicles because ESC first became available in 1997 in luxury vehicles such as Mercedes-Benz and BMW.

NHTSA has now updated and modified its 2004 report, extending it to model year 1997-2004 vehicles – and to calendar year 2004 for the FARS analysis and calendar year 2003 for the State data analysis. Nevertheless, even as of 2004, a large proportion of the vehicles equipped with ESC were still luxury vehicles. Moreover, only passenger cars and SUVs had been equipped with ESC – no pickup trucks or minivans.

The FARS database included fatal crash involvements from calendar years 1997 to 2004. The State databases included crash cases from California (2001-2003), Florida (1997-2003), Illinois (1997-2002), Kentucky (1997-2002), Missouri (1997-2003), Pennsylvania (1997-2001, 2003), and Wisconsin (1997-2003).

The basic analytical approach was to estimate the reduction of crash involvements of the types that are most likely to have benefited from ESC – relative to a control group of other types of crashes where ESC is unlikely to have made a difference in the vehicle's involvement. Crash involvements in which a vehicle 1) was stopped, parked, backing up, or entering/leaving a parking space prior to the crash, 2) traveled at a speed less than 10 mph, 3) was struck in the rear by another vehicle, or 4) was a non-culpable party in a multi-vehicle crash on a dry road, were considered the **control group** (non-relevant involvements) – because ESC would in almost all cases not have prevented the crash. The types of crash involvements where ESC would likely or at least possibly have an effect are:

- Single-vehicle crashes in which a vehicle ran off the road and then hit a fixed object and/or rolled over.
- Involvements as a culpable party in a multi-vehicle crash on a dry or wet road.
- Collisions with pedestrians, bicycles, or animals.

The principal findings and conclusions of the statistical analyses are the following:

RUN-OFF-ROAD CRASHES

- ESC reduced involvements in all types of single-vehicle run-off-road crashes the following percentages:

	Crash reduction by ESC (%)	
	Cars	LTVs
Fatal crash involvements	36	70
Police-reported crash involvements	45	72

- All four of these reductions are statistically significant.

SINGLE-VEHICLE CRASH REDUCTION

- ESC reduced all single-vehicle involvements (excluding pedestrian, bicycle, animal crashes) by the following percentages:

	Crash reduction by ESC (%)	
	Cars	LTVs
Fatal single-vehicle crash involvements	36	63
Police-reported single-vehicle crash involvements	26	48

- All four reductions are statistically significant.

ROLLOVER CRASHES

- ESC was especially effective in preventing single-vehicle first-event rollovers.

	Crash reduction by ESC (%)	
	Cars	LTVs
Fatal rollovers	70	88
Police-reported rollovers	64	85

- All four of these reductions are statistically significant.

CULPABLE INVOLVEMENTS IN MULTI-VEHICLE CRASHES

- ESC likely reduces involvements as a culpable party in multi-vehicle crashes.

	Crash reduction by ESC (%)	
	Cars	LTVs
Fatal culpable multi-vehicle crash involvements	19	34
Police-reported culpable multi-vehicle crash involvements	13	16

- Only the reduction in fatal crash involvements in passenger cars is <u>not</u> statistically significant.

COLLISIONS WITH PEDESTRIANS – BICYCLES – ANIMALS

- There are no consistently significant results in either direction for crashes that involve pedestrians, bicycles, or animals

	Crash reduction by ESC (%)	
	<u>Cars</u>	<u>LTVs</u>
Fatal pedestrian, bicycle, animal crashes	-36	-6
Police-reported pedestrian, bicycle, animal crashes	26	-11

- Only the reduction in police-reported crash involvements in passenger cars is statistically significant. We will continue to monitor the effect of ESC on this particular type of crash involvements in the future – because we do not have enough data (at the moment) for any conclusions.

OVERALL CRASH REDUCTION

- ESC reduced all crash involvements by the following percentages:

	Crash reduction by ESC (%)	
	<u>Cars</u>	<u>LTVs</u>
All fatal crash involvements	14	28
All police-reported crash involvements	8	10

- Only the reduction in fatal crash involvements in passenger cars is not statistically significant.

4-CHANNEL VERSUS 2-CHANNEL ESC SYSTEMS IN PASSENGER CARS

The passenger car sample includes certain make-models that had 2-channel ESC systems and others that had 4-channel systems. Separate analyses were performed to analyze the difference in effectiveness (if any) between 2-channel and 4-channel systems. We found:

<u>Greater</u> <u>fatal</u> run-off-road reduction with <u>2-channel systems</u> – but the larger observed fatality reduction with 2-channel systems is not statistically significantly different from the observed reduction with 4-channel systems. The reductions were most certainly influenced by the small samples.
<u>Larger</u> reduction with <u>4-channel systems</u> when police-reported crash involvements were included (as expected with larger samples). The larger reduction with 4-channel systems in all run-off-road involvements (mostly non-fatal crashes) is statistically significant.

BACKGROUND

Automotive braking technologies have evolved from very simple systems (i.e., block brakes) to more sophisticated systems (i.e., cable-operated four-wheel brakes, hydraulic four-wheel brakes, drum brakes, disc brakes with front-rear split, etc.). Today, drivers rely on much more technologically-advanced systems to help them not only to decelerate and accelerate but also to stabilize their vehicles while in motion, such as:

Antilock Brake Systems (ABS) are the first of a series of three braking technology developments. They are four-wheel systems that prevent wheel lock-up by automatically modulating the brake pressure when the driver makes an emergency stop.

Traction Control Systems (TCS) are the second technology. They deal specifically with front-to-back loss of friction between the vehicle's tires and the road surface during *acceleration.*

Electronic Stability Control (ESC) systems are another important technology evolving from and incorporating the first two technologies – ABS and TCS – with additional capabilities. They are stability enhancement systems designed to improve vehicles' lateral stability by electronically detecting and automatically assisting drivers in dangerous situations (e.g., understeer and oversteer) and under unfavorable conditions (e.g., rain, snow, sleet, ice). ESC systems have sensors that monitor the speed, the steering wheel angle, the yaw rate, and the lateral acceleration of the vehicle. Data from the sensors are used to compare a driver's intended course with the vehicle's actual movement to detect when a driver is about to lose control of a vehicle and automatically intervene in split seconds by applying the brakes to individual wheels and possibly reducing engine torque to provide stability and help the driver stay on course. For example, if an ESC system detects that the rear wheels have begun to slide to the right and the vehicle is yawing counter-clockwise, it may momentarily brake the right front wheel, imparting a clockwise torque to counteract the excessive counterclockwise yaw and stabilize the vehicle. Depending on the driving situation, these brake interventions may also be used to slow down the vehicle to a speed more appropriate for the operating conditions.

The reasons ESC appears to provide safety benefits are twofold: (1) it can anticipate situations leading up to some loss-of-control crashes before they occur, and (2) it has the capability to mitigate these crashes via automatic intervention. Hence, the potential benefit should be primarily a reduction of single-vehicle crashes that involve losing control and running off the road. These crashes include rollovers and collisions with fixed objects.

In a first published study[1] analyzing the effectiveness of ESC on three Toyota passenger car make-models, Aga and Okada reported in 2003, a 36 percent reduction in single-vehicle crash rates (single-vehicle crashes per 10,000 vehicles per year) when they compared make-model vehicles with ESC and those without ESC. The study also

---

[1] Aga, M. and Okada, A. (2003) Analysis of Vehicle Stability Control (VSC)'s Effectiveness from Accident Data, Paper Number 541, *Proceedings of the 18th International Technical Conference on the Enhanced Safety of Vehicles.*

showed a reduction (28 percent) in the rates of head-on collisions in studied vehicles that are equipped with ESC relative to those that are not equipped with ESC.

In 2004, the National Highway Traffic Safety Administration (NHTSA) initiated an evaluation to assess the effectiveness of ESC in reducing single-vehicle crashes in various domestic and imported passenger cars and Sport Utility Vehicles (SUVs). The preliminary results from that study[2] showed that ESC is highly effective in reducing single-vehicle crashes. In fact, single-vehicle crashes were reduced by 35 percent in passenger cars, and 67 percent in SUVs. Similarly, fatal single-vehicle crashes were reduced by 30 percent in passenger cars, and 63 percent in SUVs.

In a study published in 2004 by the Insurance Institute for Highway Safety (IIHS)[3], Farmer compared per vehicle crash involvement rates of vehicles with model years that had ESC as standard equipment with identical make-model year vehicles that did not. Farmer found that ESC reduced single-vehicle crash involvement risk by approximately 41 percent and reduced single-vehicle injury risk by the same amount. Overall, crash involvement risk was reduced by 7 percent for all crashes and 9 percent for injury crashes. For fatal crash involvements, it was estimated that ESC reduced single-vehicle crash involvement risk by 56 percent and overall fatal crash involvement risk by 34 percent. The effect of ESC on multi-vehicle crashes was found to be minimal and not statistically significant. Unlike the NHTSA evaluation, this study combined the results for passenger cars and SUVs.

In another study published in 2004[4], researchers in Sweden (Tingvall et al.) reported a 32 percent crash reduction with ESC on wet roads and a 38 percent reduction on roads covered with ice and snow. For all crashes except rear-end impacts, the study found an overall 22 percent reduction in crashes involving vehicles with ESC versus those without ESC. In this study, rear-end crashes on dry surfaces, assumed to be unaffected by the presence of ESC, were used as the control group. The data set used in the analysis consists of Swedish police reported crashes where at least one occupant was injured. In 2006, Tingvall et al. conducted a follow-up study analyzing the effectiveness of ESC specifically by injury severity.[5] For serious and fatal loss-of-control type crashes, it was estimated that ESC reduced those types of crashes by 56 percent on wet roads and 49 percent on roads that were covered with ice and snow.

---

[2] Dang, J. (2004) *Preliminary Results Analyzing the Effectiveness of Electronic Stability Control (ESC) Systems*, NHTSA Evaluation Note No. DOT HS 809 790, Washington, D.C.

[3] Farmer, C. (2004) Effect of Electronic Stability Control on Automobile Crash Risk, *Traffic Injury Prevention*, Vol 5, pp. 317-325.

[4] Lie, A., Tingvall, C., Krafft, M., and Kullgren, A. (2004) The Effectiveness of ESP (Electronic Stability Program) in Reducing Real Life Accidents, *Traffic Injury Prevention*, Vol 5, pp. 37-41.

[5] Lie, A., Tingvall, C., Krafft, M., and Kullgren, A. (2006) The Effectiveness of ESC (Electronic Stability Control) in Reducing Real Life Crashes and Injuries, *Traffic Injury Prevention*, Vol 7, pp. 38-43.

The latest study from the University of Michigan Transportation Research Institute[6] confirmed the results of earlier studies worldwide – ESC is highly effective in preventing single-vehicle crashes, especially rollovers in SUVs. Similar to the NHTSA evaluation, this study analyzed the effectiveness of ESC separately for passenger cars and SUVs. The study found that for single-vehicle crashes, ESC reduced the risk of a fatal crash involvement by 31 percent for passenger cars and 50 percent for SUVs. These analyses were repeated, for both passenger cars and SUVs, after restricting the data to vehicles that were three years old or newer at the time of the crash – to control for the effect (if any) of the age of the vehicle. The results showed that the vehicle age factor did not compromise the significant reductions in the risk of single-vehicle crash involvements associated with ESC-equipped vehicles. The risks of fatal rollover involvements were also decreased for passenger cars and SUVs – 40 percent and 73 percent, respectively. For the analysis of fatal crash involvements, multi-vehicle crash involvements were used as the control group. The study also analyzed a nationally representative probability sample of crashes that included mostly non-fatal crashes. It found significant reductions in run-off-road crashes for passenger cars and SUVs – 55 percent and 70 percent, respectively. For the analysis of these mostly non-fatal crashes, struck vehicles involved in rear-end crashes were used as the control group.

The 2004 NHTSA study used crash data from five States from calendar years 1997 to 2002 and Fatality Analysis Reporting System (FARS) data from calendar years 1997 to 2003, because the study was limited to the years when ESC was offered as standard equipment on certain vehicle models. Mercedes-Benz and BMW were the first two manufacturers that installed ESC as standard equipment in certain models in 1997 and in all models by 2000 and 2001, respectively. Their vehicles constituted 61 percent of the passenger car sample used in the 2004 study. The passenger car sample also includes some luxury GM cars, which constituted 23 percent of the sample, and a few luxury cars from other manufacturers. As for the sample used in the analysis for SUVs, Toyota and Lexus models constituted 78 percent of that sample. In other words, the data samples used in the 2004 study are limited to mostly imported luxury vehicles, and thus, they are not well represented across the entire fleet.

Sales of vehicles equipped with ESC are gradually increasing as shown in Table 1. In 2003, merely 10 percent of that year's vehicle sales are from vehicles that had ESC. By 2006, nearly <u>one-third</u> of the new vehicles are expected to have ESC. Thus, future analysis data samples should consist of a more representative cross-section of the fleet that will include not only non-luxury vehicles but also a wider variety of manufacturers. The availability of such data is likely to take at least a few more years. Until then, NHTSA will continue to evaluate the effectiveness of ESC with limited analysis samples of crash data from selected make-models of vehicles that had ESC and earlier versions of similar make-models that did not. The make-models used in this study consist of not only those used in the 2004 study but also other make-models and the model years extended to one more year.

---

[6] Green, P. and Woodrooffe, J. (2006) *The Effectiveness of Electronic Stability Control on Motor Vehicle Crash Prevention*, Report Number UMTRI-2006-12, University of Michigan Transportation Research Institute, Ann Arbor, MI.

Table 1: **Percent of the Vehicle Sales that are from Vehicles Equipped with ESC By Vehicle Group and Model Year**

Vehicle Group	Model Year 2003	Model Year 2004	Model Year 2005 (expected)	Model Year 2006 (expected)
Domestic cars*	3.5%	7.7%	8.2%	10.0%
Domestic trucks*	4.9%	10.2%	15.2%	26.6%
Imported cars	36.6%	45.5%	40.5%	54.3%
Imported trucks	31.8%	48.4%	63.4%	73.5%
All new vehicles	**9.7%**	**15.9%**	**19.2%**	**28.6%**

* Includes transplants

With one more year of crash data recently available, NHTSA has embarked on this update and modification of its 2004 analysis, extending it to calendar year 2004 for the FARS analysis and calendar year 2003 for the State data analysis. The primary objective of this study is to assess the effectiveness of ESC in reducing crashes, specifically crashes where ESC is likely to have made a difference in the vehicle's involvement, while controlling for others in which ESC is unlikely to have been a factor. The study consists of a series of analyses of crash data of various domestic and imported passenger cars and LTVs (light trucks and vans, including pickup trucks, SUVs, minivans, and full-size vans with Gross Vehicle Weight Rating up to 10,000 pounds) from model years 1997 to 2004. As its principal analytic tool, the agency utilized **2x2 contingency tables** of crash data from specific make-models equipped with ESC versus earlier versions of similar make-models not equipped with ESC, using non-relevant crash involvements as a control group. As a check, a logistic regression analysis of the effectiveness of ESC in reducing relevant crash involvements (specifically single-vehicle crashes which involve a vehicle running off the road, while controlling for other non-relevant crash involvements) was also performed using FARS data in order to adjust for the imbalances in various make-models of vehicles with ESC versus those of similar make-models without ESC and to account for the confounding effects from other external factors. We also looked separately at crash reduction with 2-channel and 4-channel ESC systems. The reason for doing this is that GM make-models used in the analyses (except for the Corvette) all had 2-channel ESC systems, whereas non-GM make-models all had 4-channel systems.

The next several sections discuss in detail: (1) the availability of ESC (particularly, which make-model passenger cars and LTVs had ESC and which did not); (2) the analysis databases (specifically, what and how various data files were obtained and used in the analysis); and (3) the methods of analyzing crash data and estimating the effectiveness of ESC in reducing crashes – followed by the results from a series of statistical analyses of crash data. A summary discussion of the findings concludes the report.

ESC INFORMATION

Due to the fact that (1) ESC was first introduced in the U.S. on certain luxury vehicles in 1997 and (2) it was not until recently that ESC was offered on non-luxury vehicles, information is not readily obtainable for earlier model years. For example, ESC was not listed in Buying a Safer Car[7] until 2001, and Ward's Automotive Yearbook[8] did not begin reporting ESC information in its yearbook until 2004. The yearly-published Buying a Safer Car booklet lists safety feature information for the current year's production of vehicles, whereas Ward's yearbook contains a list of factory-installed equipment for the previous year's production of vehicles. For instance, Ward's Automotive Yearbook 2001 contains a list of factory-installed equipment for vehicles produced in calendar year 2000, whereas Buying a Safer Car 2001 lists safety information for vehicles produced in calendar year 2001. Hence, information such as the availability of ESC in vehicles with model years prior to 2000 had to be obtained elsewhere (e.g., www.edmunds.com, www.cars.com). Furthermore, since ESC is a fairly new technology for the U.S. market, information on its availability can vary from one source to another. In such cases, information had to be verified with the manufacturers for accuracy.

For the purpose of analysis, make-models of passenger cars and LTVs that are equipped with ESC as standard equipment were compared with earlier versions of similar make-models. Thus, only vehicles in which 1) the percentage of factory-installed equipment for "stability control" systems increased from 0 to 100 percent (Wards' Automotive Yearbook) or 2) ESC is listed as standard in all vehicles in that model line for particular model years and not available at all for previous model years (Buying a Safer Car Booklet, etc.) were included in the analysis. Vehicles with ESC as optional equipment were excluded from the analysis because we could not determine (from the VIN) which vehicles had ESC and which did not.[9]

Tables 2 through 4 show ESC availability on selected make-model passenger cars. In some cases, models that had factory-installed ESC as a standard feature were compared with earlier versions of _identical_ models because those models remained the same (e.g. same chassis generation) during the pre- and post-ESC years – as shown in Table 2. In other words, for each vehicle listed in Table 2, exactly the same model – before and after ESC was introduced – was used for comparison. The conversion from not having ESC at all to having ESC as standard equipment should occur in consecutive model years as shown in most models except for the Chevrolet Corvette. ESC was introduced as optional equipment on the Chevrolet Corvette for model years 1998-2000. Those model years were eliminated from the study.

---

[7] This book is published yearly by NHTSA.

[8] This book is published yearly by Ward's Communication, Inc.

[9] We did examine those make-models that had ESC as optional equipment, under the assumption that all these vehicles did not have ESC, and these additional data did not make a big difference in the overall effectiveness.

Table 2: *"Identical" Make-Model Comparison of Vehicles (except Mercedes-Benz) In Model Years - With ESC and Without ESC - for Passenger Cars*

Make Model	Model Years with No ESC	Model Years with ESC (Standard)
Acura 3.5RL	1997-1999	2000-2004
Audi A6 (2.7L) Sedan	2000	2001-2004
Audi TT (1.8L) 180hp Coupe	2000	2001-2004
BMW 740I	1997	1998-2001
BMW 740IL Sedan	1997	1998-2001
Buick Park Avenue Ultra	1997-1999	2000-2004
Cadillac DeVille Concours	1997	1998-1999
Chevrolet Corvette	1997	2001-2004
Lexus GS400	1998	1999-2000
Lexus LS400	1997-1998	1999-2000
Saab 9-5 Aero Sedan/Sport Wagon	2000-2001	2002-2004

In other cases, similar make-models (except for the Mercedes-Benz vehicles which will be discussed separately) were used for the comparison since the original models were redesigned, not available, or not of the same chassis generation during the transitional years (Table 3). For instance, the 2000 Audi A4 (1.8L), 2000 A4 (1.8L) Quattro, and 2000 A4 (1.8L) Avant Quattro models were not of the same chassis generation as their 2002-2004 models. ESC was offered as optional feature on the model year 2001 for those vehicles, and thus, that model year was not included in the study. The 2004 Audi A8 4.2L Quattro was not available; thus, the 2004 A8 L was included in the analysis instead. The 1997 and 1998-2004 Cadillac Seville SLS and STS models were also on a different chassis, and so did the 1997 and 1998-2004 Lexus GS300, 1997-1999 and 2001-2003 Oldsmobile Aurora (4.0L), and 1999 and 2002-2004 Volkswagen GTI VR6 models. As previously mentioned and currently illustrated in Table 3, BMW installed ESC in certain 5, 7, and 8 series models as early as 1997 and had made it standard in all their models by 2001, but the installation did not necessarily occur in the same year for all the sub-series of a make-model. For example, the 3 series coupe and sedan models had ESC in 2000 whereas the 3 series convertible had it in 2001. Therefore, comparison of vehicles with ESC and those without ESC was done at the sub-series level. Furthermore, some manufacturers often changed their sub-series models; thus, it is very difficult to make an exact model-to-model comparison since the analysis databases consist of crash data of various make-models ranging from model years 1997-2003 for the State data analysis and 1997-2004 for the FARS data analysis. In such cases, the best approach would be to compare similar sub-series models as shown in Table 3. Since BMW's 3 and 5 series models were completely redesigned and/or grouped into different sub-series at some point during model years 1997-2004, different versions of the 3 and 5 series were used depending on which model years had ESC and which did not. Similar to BMW, General Motors changed their Pontiac Bonneville SSE model to SSEi in 2000 and installed ESC in the new model as well. The reason that model year 2000 was not included in the analysis for the BMW 3 series convertible is because that model was not available in 2000. Model year 1999 was excluded for the BMW 525i and 530i sedan for the same reason. The BMW 528i sedan was available in 1999; however, it did not include ESC as standard feature – only as an option. Hence, the 1999 BMW 528i, 525i, and 530i sedans were not included in the study. For the Volkswagen Passat, the 2000 GLS and GLX models were compared with the 2004 GLX model because the 2003 model year for both make-models had ESC as an option. The 2004 GLS model had optional ESC as well. Likewise, the 2002 Volkswagen New Beetle (1.8L) was compared with the 2002-2004 Turbo S models because the model years 2003-2004 of the former had optional ESC.

Table 3: "Similar" Make-Model Comparison of Vehicles (except Mercedes-Benz) In Model Years - With ESC and Without ESC - for Passenger Cars

Make Model	Model Years with No ESC	Make Model	Model Years with ESC (Standard)
Audi A4 (1.8L) Sedan	2000	Audi A4 (1.8L) Sedan	2002-2004
Audi A4 (1.8L) Quattro Sedan	2000	Audi A4 (1.8L) Quattro Sedan	2002-2004
Audi A4 (1.8L) Avant Quattro Wagon	2000	Audi A4 (1.8L) Avant Quattro Wagon	2002-2004
Audi A8 (4.2L) Quattro Sedan	1998-1999	Audi A8 (4.2L) Quattro Sedan Audi A8 L Sedan	2001-2003 2004
BMW 318/323/325/328is/M3 Coupe (6cyl)	1997-1999	BMW 323CI/325Ci/328Ci/330/M3CI Coupe (6cyl)	2000-2004
BMW 323iC/328iC/M3 Convertible	1997-1999	BMW 325Ci/330Ci/M3Ci Convertible	2001-2004
BMW 318/323i/328iSedan; BMW 318TI 2Dr.	1997-1999	BMW 323/325i/328i/330i/Xi Sedan	2000-2004
BMW 528i Sedan	1997-1998	BMW 528i/525i/530i Sedan	2000-2004
BMW 540i (AT/MT) Sedan	1997	BMW 540i (AT) Sedan	1998
BMW 540i (MT) Sedan	1998	BMW 540i/545i (AT/MT) Sedan	1999-2004
BMW Z3 (2.8L)	1997-1999	BMW Z3 (2.3L,2.5L,2.8L,3.0L)	2000-2002
Cadillac Seville SLS	1997-1998	Cadillac Seville SLS	1999-2004
Cadillac Seville STS	1997-1998	Cadillac Seville STS	1999-2004
Lexus GS300	1997-1998	Lexus GS300	1999-2004
Oldsmobile Aurora (4.0L)	1997-1999	Oldsmobile Aurora (4.0L)	2001-2003
Pontiac Bonneville SSE	1997-1999	Pontiac Bonneville SSEi	2000-2004
Volkswagen GTI VR6	1999	Volkswagen GTI VR6	2002-2004
Volkswagen Passat GLS/GLX Sedan/Wagon	2002	Volkswagen Passat GLX Sedan/Wagon	2004
Volkswagen New Beetle (1.8L)	2002	Volkswagen New Beetle Turbo S (1.8L)	2002-2004

Like BMW, Mercedes-Benz began installing ESC in certain sub-series of the S class model in 1997 and had made it a standard feature in all their make-models (except one) by 2000. For this reason, not all the sub-series (even within the same make-model – i.e. "class") had ESC in the same model year. Furthermore, most sub-series were changed during calendar years 1997-2004 – either redesigned or replaced. Table 4 lists all the available Mercedes-Benz models and sub-series from 1997 to 2004, but only with the model years that had ESC as standard equipment and those that did not. In Table 4, various sub-series within the Mercedes-Benz C, E, and S class that are equipped with ESC were compared with earlier versions of <u>similar</u> sub-series, for reasons already discussed in the previous section. Unlike other manufacturers, quite a few Mercedes-Benz sub-series had ESC as optional equipment for one, two, or even three years; hence, those model years were not included in the analysis. For example, the C280, E300DT sedans and the CL500C coupe all had ESC as an optional feature from 1998 to 1999.

<u>Table 4</u>: *Mercedes-Benz Make-Model Comparison of Model Years With ESC and Without ESC - for Passenger Cars*

Make Model	No ESC	Make Model	ESC (Standard)
C36AMG	1997	C43 Sedan	1998-2000
C220, C230 Sedan	1997-1999	C230ML Sedan	2000
C280 Sedan	1997	C280 Sedan	2000
		C Sedan	2001-2004
		CL Coupe	2000-2004
		CL600 Coupe	1998-1999
		CLK55AMG Coupe/Cabriolet	2001-2002
		CLK320 Coupe/Cabriolet	2000-2003
		CLK320/CLK500 Coupe/Cabriolet	2003-2004
		CLK430 Coupe/Cabriolet	1998-2003
E300D Sedan	1997	E320 Sedan/Wagon	2000-2003
E320 Sedan	1997	E320W/E500 Sedan	2003-2004
		E430/E55 AMG Sedan	1999-2002
S320 SWB Sedan	1997-1999	S430/S500/S600 Sedan	2000-2004
S320 LWB Sedan	1997	S600 Sedan	1997-1999
SL320 Roadster	1997	S600 Coupe	1997
SLK230 Kompressor	1998-2000	SL500 Roadster	1999-2002
		SL500/SL55 Roadster	2003-2004
		SL600 Roadster	1997-2002
		SLK230/SLK320/SLK32AMG	2001-2004

Tables 5 through 7 show the availability of ESC by make-model for selected LTVs. As shown in these tables, the make-models used in the analysis consist of SUVs and passenger vans – because most manufacturers had not yet installed ESC in their pickup and minivan models as of 2004 – the most recent and available fatal crash data used in this study. With the exception of the Infiniti, Land Rover, Toyota 4 Runner, Mercedes-Benz, and one Lexus model, none of the models used in the analysis went through a redesign in the year that ESC became standard equipment or even in the year immediately before or after it (Table 5). Thus, we were able to compare several model years of <u>identical</u> make-models for the vehicles with ESC versus those without ESC. Furthermore, these vehicles went from not having ESC at all in one year to having ESC as standard equipment in the following year.

Table 5: *"Identical" Make-Model Comparison of LTVs In Model Years With ESC and Without ESC*

Make Model	Model Years with No ESC	Model Years with ESC (Standard)
Acura MDX	2001-2002	2003-2004
Cadillac Escalade 2WD	2002	2003-2004
Chevy Express 3500 Extended Wheelbase	2003	2004
GMC Yukon Denali AWD	2001-2002	2003-2004
GMC Yukon Denali XL AWD	2001-2002	2003-2004
GMC Savana G3500 Extended Wheelbase RWD	2003	2004
Lexus LX470	1999	2000-2004
Mitsubishi Montero Limited	2001-2002	2003-2004
Toyota Land Cruiser	1999-2000	2001-2004
Toyota RAV4 4x2	2003	2004
Toyota RAV4 4x4	2003	2004

As for the Infiniti vehicles listed in Table 6, since there are no previous versions of the FX35 AWD, FX45AWD, and QX56 models – which had ESC starting in 2003 – the model years 2002-2003 of the QX4 model were used as comparison vehicles that did not have ESC. The 2001-2002 and 2003-2004 Land Rover Range Rover models were not of the same chassis generation. The Toyota 4Runner had a design change in 2003 for both the 4x2 and 4x4 models. The static stability factor was higher on the 2003-04 4Runners. Hence, the later model-year vehicles (post 2002 model year) are more stable than the earlier model-year vehicles. Thus, one would expect the 2003-04 4Runners to be involved in fewer rollover crashes than the earlier models, even if the newer models are hypothetically not equipped with ESC. For that reason, the 2003-04 model years were not included in the analysis.[10] Lexus changed their RX300 model to RX330 in 2004, the new model was included in the analysis.

The Mercedes-Benz ML320 – Mercedes' first ever U.S. SUV model – came on the market in 1998 but was not equipped with ESC. A year later, Mercedes-Benz had made

---

[10] Charles Farmer, in his peer review of this report, recommended the exclusion of 2003-2004 Toyota 4Runners.

ESC standard equipment in all their SUV models.   We could have included only the
ML320 model in the analysis – since the model was still available in 2004; however, we
decided to include other Mercedes SUV models in the analysis to increase our sample
size.

Table 6: *"Similar" Make-Model Comparison of LTVs In Model Years*
With ESC and Without ESC

Make Model	Model Years with No ESC	Make Model	Model Years with ESC (Standard)
Infiniti QX4 2WD	2002-2003	Infiniti FX35 AWD	2003-2004
Infiniti QX4 4WD	2002-2003	Infiniti FX45 AWD	2003-2004
		Infiniti QX56 4x2	2004
		Infiniti QX56 4x4	2004
Land Rover Range Rover	2001-2002	Land Rover Range Rover	2003-2004
Lexus RX300	1999-2000	Lexus RX300	2001-2003
		Lexus RX330	2004
Toyota 4Runner 4x2	1999-2000	Toyota 4Runner 4x2	2001-2002
Toyota 4Runner 4x4	1999-2000	Toyota 4Runner 4x4	2001-2002

Table 7: *"Similar" Make-Model Comparison of Mercedes-Benz LTVs In Model Years*
With ESC and Without ESC

Make Model	Model Years with No ESC	Make Model	Model Years without ESC (Standard)
ML320	1998	ML320	1999-2004
		ML350	2003-2004
		ML430	1999, 2001-2002
		ML500	2002-2004
		ML55	2000, 2004

CRASH DATA (STATE AND FARS)

Once specific make, model, and model year of vehicles – equipped with ESC and those
not equipped with ESC – were selected for passenger cars and LTVs, those vehicles can
then be identified in the crash data for analysis. In general, a large sample of crash-
involved cars from the selected models is desirable to statistically estimate the effect of
ESC in reducing crashes. The best source of data available to NHTSA for this analysis is
the State data files maintained by the agency's National Center for Statistics and Analysis
(NCSA). NCSA currently receives crash data from 28 States and maintains these data
files for calendar years 1989 and onward. Since ESC is currently not required safety
equipment, its presence in crash-involved vehicles is not encoded as a separate field in

the police reports. However, we can identify the make, model, and model year of each crash-involved vehicle from the State data files by decoding the Vehicle Identification Number (VIN) – if such information is available.

Twenty-one States were excluded from the analysis for various reasons. For instance, Alabama, Arkansas, Colorado, Delaware, Indiana, Minnesota, Montana, South Carolina, Texas, and Virginia do not have VIN information in their files. Georgia, Michigan, North Carolina, New York, and Washington were excluded because NHTSA does not have their files beyond calendar year 2000, a critical year in the analysis because the majority of our passenger car sample consists of Mercedes-Benz and BMW vehicles – the first two manufacturers that offered ESC on all their models by 2000 and 2001, respectively. Thus, without the post 2000 data, our data sample would consist of mostly crashes from non-ESC vehicles. Kansas, New Mexico, and Ohio were not uniform in reporting VIN, having a low percentage of VIN in some years and a high percentage in others. Utah and Wyoming are States with small numbers of crashes. Maryland data in nearly half of the reported cases do not indicate the contributing circumstances for the vehicles involved in the crashes. We must have this information to 1) determine which vehicle was responsible for the crash and which was not and 2) classify each vehicle's involvement as either the control group or response group (relevant) involvement.

As a result, data from seven populous States (California, Florida, Illinois, Kentucky, Missouri, Pennsylvania, and Wisconsin) were used in the analysis. These States (except California) consistently have a high percentage of VIN information in their data files (Table 8). California does not have VIN information in its data files. Thus, we were not able to identify (from the VIN) specific make-models that had ESC and those that did not, but we included California in the analysis because it is a large State with large numbers of crash data. We know that Mercedes-Benz first installed ESC in certain S class models and had made it standard equipment in all their passenger car models by 2001. Mercedes-Benz did have a few models that had ESC as optional equipment prior to 2001. As previously mentioned, we did examine vehicles with ESC as optional equipment under the assumptions that those vehicles did not have ESC, and the initial results indicated that the additional data did not make a big difference in the overall effectiveness. Thus, for the analysis of crash data in California, we assumed that all Mercedes-Benz passenger car models prior to 2000 did not have ESC at all. As for their SUV models, Mercedes-Benz installed ESC in all their models from model years 1999 and onward. Thus, only the 1998 Mercedes-Benz SUV models did not have ESC. As a result, in the California analysis, we were able to analyze crashes that involved only Mercedes-Benz vehicles – because we were able to identify those vehicles using the variables MAKE, MOD_YR (model year), and CHP_TYP (California Highway Patrol vehicle type – to differentiate cars from SUVs).

Although the number of vehicles involved in fatal crashes is relatively small, ESC is believed to be highly effective in reducing fatal crashes. If so, even a relatively small sample size will suffice for statistically significant effects. Currently available crash data from Fatality Analysis Reporting System (FARS) from calendar years 1997-2004 were used for the fatal crash analyses. FARS data files consistently have high percentage of

VIN information in their data files (97% or better). Thus, the make, model, and model year of fatal crash-involved vehicles can be decoded from the VIN information.

Table 8: Percentage of VIN Information by State and Calendar Year

State	Calendar Year 1997	Calendar Year 1998	Calendar Year 1999	Calendar Year 2000	Calendar Year 2001	Calendar Year 2002	Calendar Year 2003
Florida	90%	90%	90%	89%	88%	90%	90%
Illinois	86%	88%	87%	77%	80%	87%	-
Kentucky	95%	95%	92%	97%	97%	97%	-
Missouri	84%	87%	87%	88%	87%	93%	93%
Pennsylvania	93%	94%	94%	94%	95%	-	95%
Wisconsin	91%	91%	90%	91%	90%	90%	90%

Since NCSA typically receives crash data from the States two years after or even longer in some States, our State data analyses consist of crash data from calendar years 1997-2003 for Florida, 1997-2002 for Illinois, 1997-2002 for Kentucky, 1997-2001 and 2003 for Pennsylvania, and 1997-2003 for Wisconsin – Table 9. Currently, the 2003 Illinois, 2003 Kentucky, and 2002 Pennsylvania crash files are not available. Only data from calendar years 2001-2003 were included in the California analysis because we could not distinguish (from the "vehicle type" variable in the data files prior to 2001) between crashes involving passenger cars and those involving SUVs. Post-2000 California data files differentiate passenger car and SUVs crash involvements. As for the fatal crash analyses, FARS data files are generally available one year after. Thus, our FARS data analysis consists of fatal crash data from calendar years 1997-2004.

Table 9: Calendar Years of Crash Data By State

State	Calendar Years
California	2001-2003
Florida	1997-2003
Illinois	1997-2002
Kentucky	1997-2002
Missouri	1997-2003
Pennsylvania	1997-2001, 2003
Wisconsin	1997-2003

ANALYSIS DATABASES

A) VIN Decode

The analysis databases are crash files initially obtained from the FARS and State data files and then decoded from the VIN to include only the selected make-models listed in

12

Tables 2-7. In other words, in Florida, Illinois, Kentucky, Missouri, Pennsylvania, and Wisconsin State data files, vehicle make, model, and model year were decoded from the variable VIN and matched with the VIN information obtained from the Passenger Vehicle Identification Manual[11] for the selected make-models. To ensure that there are no discrepancies in reporting VIN information of crash-involved vehicles, the decoded model year from the variable VIN had to match with the variable MOD_YR (model year) taken directly from the State data files. Since California does not have the VIN variable, the selected Mercedes-Benz make-models were obtained by using the variable MAKE in the data files. Like other States, California does have the variable MOD_YR in its data files; thus, the model year of crash-involved vehicles was obtained directly from that variable.

Like the State data files (except California), FARS data files include both the variables VIN and MOD_YEAR (model year). Thus, we were able to extract information that identified the vehicle make, model, and model year from the VIN variable and established an analysis database consisting of crash data of make-models listed in Tables 2-7. Also, the model year taken from the VIN variable had to correspond with the variable MOD_YEAR.

The resulting analysis databases – whether they are FARS or State files – are vehicle-oriented files, with one record for each vehicle that was involved in a crash. Since these files only contain the selected make-models from Tables 2-7, only vehicles with model years 1997-2004 for the FARS files and model years 1997-2003 for the State files were included in the study. One of the critical parameters used in the databases is the parameter that identifies the presence or absence of ESC. Each record in the databases will have a variable **ESC** with a value of 1 for a vehicle that has ESC and a value of 0 for the one that does not.

B) Crash involvements

Certain crash involvements in which a vehicle 1) was stopped, parked, backing up, or entering/leaving a parking space prior to the crash, 2) traveled at a speed less than 10 miles per hour (mph), 3) struck in the rear by another vehicle, or 4) was a non-culpable party in a multi-vehicle crash on a dry road, were considered the **control group** (non-relevant involvements) – because ESC would not have prevented the crash. Since ESC has the ability to detect when a driver is about to lose control of a vehicle and the capability to automatically intervene to assist that driver and ultimately reduce the likelihood of a roadside departure, it may provide considerable benefits to crashes that involve a vehicle *running off the road* or *running out of the lane and hitting other vehicles*. Most run-off-road crashes are single-vehicle involvements, and these involvements include rollovers and collisions with fixed objects. Other relevant crash involvements – where ESC could be a factor in the vehicle's involvement – include collisions with one or more vehicles due to the driver's error or the vehicle's unsatisfactory performance (known as culpable involvements in multi-vehicle crashes). These crashes are relevant because they may (but not necessarily) have involved loss of

---

[11] This manual is published yearly by the National Insurance Crime Bureau (NICB).

control. Hence, these relevant involvements were included in the analysis and considered the **response group**. We also analyzed crashes that involved a pedestrian, bicycle, or animal – although the vehicles involved in these crashes were not typically at fault. We would still be interested in knowing whether or not ESC is effective in reducing these types of crashes.

An important concept in setting up a control- and response-group experiment is that not every vehicle in the response group necessarily went out of control or stood to benefit from ESC. For example, a careless driver could steer a vehicle off the road or into the path of another vehicle with full directional control. Moreover, these crash data generally do not specify if a vehicle "went out of control" or not. Instead, it is the converse that is true: if a vehicle did go out of control and get into a crash, it will be in a response group and not in the control group. The vehicles that could have benefited from ESC are a subset of the response-group involvements and are not control-group involvements. As mentioned previously, vehicles that most likely would not benefit from ESC are considered the control-group involvements – and while this group ideally should be as unaffected as possible by ESC, the effectiveness estimates are quite dependent on the choice of the control group. For example, if crashes that were affected by ESC were included in the control-group involvements, the estimate of ESC effectiveness would be lower. On the other hand, if crash involvements that were unaffected by ESC were included in the response-group, the effectiveness estimate would not be compromise as long as the set-up of a control- and response-group experiment was correct and valid.

The next several sections discuss how these crash involvements were obtained. The control-group involvements (non-relevant) will be discussed first followed by the relevant involvements – the response group.

Control group – non-relevant crash involvements

Ideally, only crash involvements where a vehicle was standing still prior to the crash, moving less than 10 miles per hour, or backing up should be considered the control group – since we are almost certain that these vehicles were moving too slowly to trigger ESC and make it a factor in the vehicle's involvement. But such fatal involvements are few in number – when compared to the relevant involvements. Thus, a larger sample of vehicles is needed for the control group. Other non-relevant involvements where ESC is unlikely to provide any benefit include situations where 1) a vehicle was struck in the rear in a rear-end collision or 2) a vehicle was involved in a multi-vehicle crash on a dry road but was not at fault. In both situations, the motion of this vehicle is not what precipitated the crash, and this vehicle definitely did not hit something else because it had gone out of control.

In FARS, vehicle "maneuver" and "travel speed" prior to the crash were used to define one portion of the control-group involvements in which a crash-involved vehicle was either at a complete stop or moving at a low speed – less then 10 miles per hour. In addition, the maneuvers "backing up", "parking", or "leaving a parking space" indicate the vehicle was moving slowly (most likely less than 10 mph) even if FARS does not specify a travel speed. The other portion of the control group includes vehicles that are considered a non-culpable party in a fatal multi-vehicle crash on a dry road. To determine whether or not a vehicle was at fault, related factors such as the driver's physical and mental condition, attitude, and driving actions as well as the condition of the vehicle and roadway (prior to the crash) were taken into consideration. These factors are coded in FARS and State data files. Essentially, a vehicle is in the control group if none of these factors were present. Multi-vehicle crash involvements will be discussed separately in the next section.

In theory, "not at fault" involvements on wet roads should be in the control group. However, given that many involvements follow loss of control, and given some uncertainty as to the accuracy of the "fault" determination, those involvements were not assigned to the control group.

Finally, in a front-to-rear collision (neither vehicle backing up), we will always assign the rear-impacted vehicle to the control group and the frontally impacting vehicle to the response group.

As for the States, every State has its own unique way of coding vehicle "maneuver", "speed", and initial "point of impact", and not every State includes these variables in their data files. Similar to FARS, contributing circumstances relating to the driver, vehicle, and roadway were used to determine which vehicle was responsible for the crash and which was not. If a State has a variable that indicates whether or not the vehicle's driver was charged with a moving violation, then we also used that variable to classify the vehicle's involvement (culpable or non-culpable) in a multi-vehicle crash. Again, not all variables are included in every State, and even the same variables are not coded exactly the same in each State. Thus, depending on what crash information is available in each State, the control group used in each of the State data analyses includes different types of non-relevant crash involvements. But at least the definitions for each of these types of involvements are made as similar as possible.

The definitions used in all the State data analyses for "stopped", "parked", "backing", and "parking" maneuvers are self-explanatory and were taken directly from the variable VEH_MAN1 in the State data files. If a State also reports the speed of a vehicle (variable SPEED) prior to the crash, then the crash-involved vehicle with a speed ranging from 0 to 10 mph would be included in the control group as well. Similar to the control group used in the FARS analysis, the multi-vehicle crash involvements had to occur on dry roads and be *non-relevant* (non-culpable) – in order for those involvements to be included in the control group for the State data analyses. The next section will discuss in detail the different types of multi-vehicle crash involvements used in the FARS and State analyses.

Classification of multi-vehicle crash involvements (control and response group)

Multi-vehicle crash involvements were classified as either culpable or non-culpable based on what information is available in the data files. In many crashes that involve two motor vehicles in transport, one vehicle can be identified as responsible for the crash, and the other one not. In some State files this is almost a hard-and-fast rule, with one culpable and one non-culpable vehicle in nearly every crash, but in other States there may be many crashes where none or sometimes even all the vehicles are judged culpable. In multi-vehicle crashes other than front-to-rear collisions, the action of the driver, the characteristic of the vehicle, the condition of the roadway, or simply the traffic violation charged were taken in account to determine which vehicle(s) (if any) contributed to the crash and which did not. In general, the vehicle's involvement – where the driver in that vehicle was distracted or under the influence of drugs and/or alcohol, drove carelessly or aggressively, failed to yield right-of-way, performed an improper turn, backing, passing, or lane change, followed too closely, disregarded a traffic signal, sign, or other traffic control, or exceeded the stated speed limit – was considered the most likely to cause the crash. Other contributing circumstances include the condition of the vehicle's tires (i.e., worn, blowout, puncture, etc.) or the malfunction of the vehicle's brake systems or the steering mechanism prior to the crash. Although certain roadway conditions (i.e., wet, slush, snow, icy, sand, dirt, oil, etc.) might have caused the driver to lose control of the vehicle and run off the road or hit another vehicle, the striking vehicle would still be considered the culpable party. Furthermore, if a crash-involved vehicle in which the driver were charged with a traffic violation, then that vehicle would be considered the offender. Other contributing factors – whether they are driver, vehicle, or roadway contributing circumstances – were not considered evidence of culpability. As an exception to this approach, in a two-vehicle front-to-rear collision, neither vehicle backing up, the collisions are identified by MAN_COLL="rear-end" on the accident level and IMPACT1 (or its equivalent) on the vehicle level: one vehicle has frontal damage and the other, rear damage. The vehicle with frontal damage (known as rear-end striking) will always be classified as the culpable party, and the one with rear damage (known as rear-end struck) will be non-culpable. The criteria for determining culpability in the FARS analysis are similar to those used in the State analysis. In addition to the ideal control-group involvements (i.e., vehicle was parked, stopped, or traveled at very slow speed), the control-group also included non-culpable involvements in multi-vehicle crashes on dry roads.

Up until now, we have discussed all control-group involvements in which a crash-involved vehicle was 1) standing still, 2) moving less than 10 miles per hour, 3) backing up, 4) parking or leaving a parked position, 5) being struck in the rear by another vehicle, or 6) non-culpable in a multi-vehicle crash on a dry road. We have also talked about certain relevant involvements (i.e., response group) where 1) a crash-involved vehicle was considered a culpable party in a multi-vehicle crash on any roads (dry or wet) and 2) ESC could play a critical role in preventing those crashes.

Categories of single-vehicle crashes

Now, let us discuss further other relevant involvements – mainly crashes that involve only one motor vehicle in transport – single-vehicle crashes. In most single-vehicle involvements – whether they are fatals or non-fatals – the case vehicle either rolled over or hit a fixed object. Those incidents will most likely occur when the vehicle <u>unintentionally</u> leaves the travel lane (runs off the road), which is an indication of a driver's error due to loss of steering and/or directional control of the vehicle, because it is very unlikely for a vehicle (even for the high center-of-gravity vehicle) to tip over or hit a fixed object while it is still in the travel lane and under control. In other single-vehicle involvements where the driver intentionally steered the vehicle away from the incoming object (i.e., pedestrian, bicycle, or animal) to avoid hitting the object but ended up hitting it, the driver is typically not at fault in this case, but the involvement is still considered relevant because ESC could perhaps play a role in the vehicle's involvement.

Run-off-road crashes involve vehicles that travel out of the lane – where they may or may not yaw out of control, leave the roadway, and eventually roll over and/or hit one or more natural or artificial objects along or off the roadside – such as trees, guardrails, embankments, etc. They can also occur on the median of a divided roadway or on the other side of a non-divided roadway. Run-off-road crashes typically involve a single motor vehicle in transport – although a moving vehicle hitting a parked vehicle could also be considered a multi-vehicle involvement. For the purpose of analysis, those involvements are treated as single-vehicle involvements.

We know from our preliminary study, as well as studies by other researchers that ESC appears to provide significant benefits related to single-vehicle crashes – especially those that are likely to have involved yawing: rollovers and side impacts with fixed objects. "Rollover" generally involves a vehicle that went out of control, started to yaw while it was still on the road, and eventually rotated 90 degrees or more, side-to-side or end-to-end, as a result of a tripping mechanism when it left the roadway. "Side impact with a fixed object" usually indicates a vehicle that left the roadway, possibly after yawing out of control, and collided with a fixed object. We were able to define these crash involvements using the "first harmful event" variable at the accident level and the "initial point of impact" variable at the vehicle level. These variables are coded in the FARS and State data files.

In the 2004 NHTSA's report, we studied ESC by looking at its effectiveness in reducing single-vehicle crashes as a whole and excluding crashes that involve a vehicle colliding with a pedestrian, bicycle, or animal. In other words, the response group (used in the 2004 study) excluded all crash involvements with a pedestrian, bicycle, or animal and included all other involvements with one motor vehicle in transport as single-vehicle crashes. In this follow-up study, we not only included pedestrian, bicycle, and animal crashes in our study but also looked at other single-vehicle crashes separately. In other words, we analyzed separately – the effectiveness of ESC in reducing 1) rollovers, 2) side impacts with fixed objects 3) other run-off-road crashes, 4) pedestrian, bicycle, and animal crashes, and 5) other single-vehicle crashes (such as collisions with parked cars, impacts with thrown or falling objects, undercarriage scrapes, or first-event fires). Rollovers, side impacts with fixed objects and other run-off-roads are considered "all run-off-road" crashes, and they are a subset of the single-vehicle involvements used in the 2004 study. "All run-off-road" and "other single-vehicle" involvements used in this study are what we have used in the 2004 study as "single-vehicle crashes" – since we excluded pedestrian, bicycle, and animal crashes in that study. However, a direct comparison of the relative effectiveness of ESC should be made with caution since the control groups used in the two studies are not the same – as already discussed in the previous section. In this study, we also included culpable involvements in multi-vehicle crashes on all roads as part of the response group – because ESC could be a factor in the vehicle's involvement. We also calculated the effect of ESC on all crash involvements, including the control-group involvements as well as the relevant involvements.

2X2 CONTINGENCY TABLE ANALYSIS

This study consists of a series of analyses of crash data from currently available State and FARS databases. Crash data from calendar years 1997-2003 from seven States as well as FARS data from calendar years 1997-2004 were used in the analysis. The analysis then compares specific make-models of passenger cars and LTVs with ESC versus earlier versions of similar make-models, using non-relevant crash involvements as a control group, essentially creating 2x2 contingency tables of crash involvements (control group and response group) of vehicles with ESC and those without ESC. A general notation of a 2x2 contingency table is shown below in Table 11.

Table 11: General Notation for a 2x2 Contingency Table

Type of crash involvement	Vehicles *Without* ESC	Vehicles W*ith* ESC	Totals
Control-group (non-relevant crashes)	a	b	a + b
Response group (relevant crashes)	c	d	c + d
Total	a + c	b + d	a + b + c + d = N

To detect if ESC is effective, we considered and tested the null hypothesis that the ratio of d to c is the same as the ratio of b to a. To calculate the effectiveness of ESC, first, we compute the ratios of relevant crash involvements to control-group involvements. Then, we compute the percentage reductions in these ratios in vehicles with ESC versus earlier versions of similar make-models without ESC. These calculations are illustrated below by the formula:

$$Effectiveness \ (\%) \ = \ \{1-[(d/c)/(b/a)]\} \times 100$$

To test for statistical significance of the effectiveness of ESC, we used the chi-square statistic (χ^2) for the 2x2 table.

For a 2x2 contingency table, the degrees of freedom are always 1. For a two-sided 95% confidence level (a conventionally accepted significance level) and degrees of freedom equal to 1, a chi-square statistic has to exceed 3.843 – which indicates that the two distributions are not the same – to reject the null hypothesis and accept our hypothesis that the ratio of relevant to control-group crashes is lower in the ESC-equipped vehicles than in the comparison vehicles without ESC. We will also accept as "statistically significant" a one-sided 95% confidence level, when chi-square exceeds 2.71. NHTSA evaluations of safety equipment customarily employ the more lenient one-sided test when there is a clear expectation that the effect of the equipment will be in the "right" direction (saving lives, preventing crashes) or, at worst, zero. It is unlikely to be negative. Only in situations where there is no realistic a priori expectation of an effect in either direction do we rely exclusively on the more stringent two-sided test.

The next several sections will show 2x2 contingency tables of different types of crash involvements (FARS and State files) for vehicles with ESC versus those with no ESC. As previously discussed, the control group includes crash involvements in which a vehicle was standing still prior to the crash, moving less than 10 mph or backing up, struck in the rear, or non-culpable party in a multi-vehicle crash on a dry road. The tables will also show different types of relevant crash involvements such as single-vehicle run-off-road, other single-vehicle collision, culpable involvement in a multi-vehicle collision, and collision with a pedestrian, bicycle, or animal. Run-off-road involvements such as side impacts with fixed objects, rollovers, and other run-off-roads will also be included in these tables. *All run-off-road involvements discussed in this study are single-vehicle involvements.* In addition, the effectiveness of ESC and the chi-square statistic of each type of relevant involvement will be calculated and illustrated in these tables. Lastly, the overall effectiveness of ESC on all crash involvements (including the culpable, non-culpable, and control-group involvements) will be shown in these tables – to demonstrate the overall effect of ESC since these relevant involvements consist of only a fraction of all the involvements. Thus, the actual effectiveness will always be less when considering all crash involvements.[12] For instance, if the effectiveness of ESC is estimated to be 30 percent for the relevant involvements (i.e., all run-off-roads, other single-vehicle collisions, collisions with pedestrians, bicycles, or animals, and all culpable multi-vehicle collisions), which are 50 percent of all crash involvements, then the effectiveness on all crash involvements is reduced to

$$(0.30 \times 0.50) \times 100 = 15 \text{ percent}$$

As previously mentioned, the control group includes not only crash involvements in which a vehicle was standing still prior to the crash, moving at a very low speed, or backing up, but also other involvements where a vehicle was considered a non-culpable party in a multi-vehicle crash on a dry road. For completeness, these tables will also show other multi-vehicle involvements for vehicles with ESC and those without ESC – so that the sum of all the involvements (control group, all run-off-road, other single-

---

[12] For the analysis of fatal crashes obtained from the FARS database, *all crash involvements* refer (in this study) to all types of fatal crash involvements. For the analysis of police-reported crashes obtained from the State data files, *all crash involvements* also refer to all types of crash involvements including not only fatal but also non-fatal crash involvements (i.e., property damage, possible injury, non-incapacitating and incapacitating injury).

vehicle, pedestrian, bicycle, animal, culpable involvements as well as other involvement in multi-vehicle crashes) equals the "total" involvements, which are shown prior to the last row in each of the tables. But the effectiveness and chi-square statistic will not be calculated for other multi-vehicle involvements because interpretation of the effectiveness estimates on those involvements is not clear – since these involvements also included the non-relevant involvements in multi-vehicle crashes on non-dry roads. As for the "total" involvements, the effectiveness calculation is also not suitable because they also include the control-group involvements.

Fatal crashes in passenger cars

Now let us first look at the effectiveness of ESC in fatal crashes in passenger cars. As illustrated in Table 12, ESC reduced fatal run-off-road crashes by 36 percent $[1 - \{(154/217)/(183/166)\}]$. The reduction is statistically significant with a chi-square statistic of 8.62 for the 2x2 table consisting of these four cells. It is also evident that fatal rollover risk is substantially lower with ESC – a statistically significant 70 percent reduction relative to the control group. Fatal side impacts with fixed objects were significantly reduced as well – as shown by the 49 percent effectiveness and a 5.89 chi-square statistic. The negative effectiveness (-36 percent) relative to the control group in pedestrian, bicycle, and animal crashes suggests possible increase, but the increase is statistically non-significant. On a more positive note, culpable involvements in multi-vehicle crashes were also decreased (19 percent), but the reduction is not statistically significant. Overall, ESC is still very beneficial (14 percent effectiveness) if all crash involvements were considered – which included culpable, non-culpable, and the control-group involvements. To calculate the overall effectiveness, the sum of all run-off-roads, other single-vehicle, pedestrian, bicycle, and animal involvements, as well as culpable and non-culpable involvements in multi-vehicle crashes must first be calculated for vehicles with ESC and those with no ESC [(154+3+69+157+76=459) and (217+6+46+176+65=510), respectively]. The effectiveness in these non-control group involvements relative to the control-group involvements can then be calculated [1-$\{(459/183)/(510/166)\}$ = 18 percent]. The effect of ESC on all fatal crash involvements for passenger cars can finally be determined by multiplying the effectiveness in non-control group involvements by the proportion of the total crash involvements that were relevant [18 x (510/676) = 14 percent]. The effect of ESC on all crash involvements is not statistically significant because the effect on all non-control group crash involvements is not.

As previously discussed, *"all run-off-road" and "other single-vehicle" involvements used in this study are what we have categorized in the 2004 study as "single-vehicle crashes" – which did not include pedestrian, bicycle, and animal crashes.* For comparison purpose (though the control groups used in the 2004 study and in this study are not the same), we also included in the analysis tables the effectiveness estimates of ESC in reducing *single-vehicle crashes except those involving pedestrians, bicycles, or animals.* The results from this study (shown in Table 12) suggest that ESC reduced fatal single-vehicle crash involvements of passenger cars by 36 percent – if pedestrian, bicycle, and animal crashes were excluded – and the reduction is statistically significant.

Table 12: Effect of ESC on Fatal Crashes in Passenger Cars
(FARS: 1997-2004)

Type of fatal crash involvement	Vehicles with no ESC	Vehicles with ESC	Effectiveness	*Chi-square Value
	Fatal Crashes	Fatal Crashes	(%)	
Control group	*166*	*183*		
All run-off-road	217	154	36	8.62
Side impact with fixed object	41	23	49	5.89
Rollover	36	12	70	12.71
Other run-off-road	140	119	23	2.50
Pedestrian/bicycle/animal	46	69	-36	1.99
Other single vehicle	6	3	55	1.28
Culpable multi-vehicle	176	157	19	1.91
Other multi-vehicle*	65	76		
All non-control group involvements	510	459	18	2.64
All crash involvements			14	
Total	676	642		
Single-vehicle crashes	223	157	36	9.04

* For the 2x2 table formed by this row and the "control group" row.

** excludes pedestrian, bicycle, animal crashes.

*** includes involvements that are not culpable but are not part of the control-group.

Fatal crashes in LTVs

For LTVs (the majority of the analysis sample consists of SUVs), ESC is highly effective in reducing fatal run-off-road crashes (70 percent effectiveness, chi-square statistic 29.90), especially rollovers (88 percent) – Table 13. In this analysis, ESC has little observed effect on crashes that involved pedestrians, bicycles, or animals. Culpable involvements in fatal multi-vehicle crashes significantly declined (34 percent effectiveness; chi-square statistic 3.64) in LTVs with ESC. The overall effectiveness in all fatal crash involvements is a positive 28 percent – which indicates a substantial benefit with ESC in reducing all fatal crashes in LTVs. Similar to the result from the 2004 study, ESC is highly effective in reducing single-vehicle crashes involving LTVs – as shown in Table 13 by the positive 63 percent effectiveness, and 23.32 chi-square statistic.

21

Table 13: Effect of ESC on Fatal Crashes in LTVs
(FARS: 1997-2004)

Type of fatal crash involvement	Vehicles with no ESC	Vehicles with ESC	Effectiveness	*Chi-square Value
	Fatal Crashes	Fatal Crashes	(%)	
Control group	*153*	*95*		
All run-off-road	191	36	70	29.90
Side impact with fixed object	15	6	36	0.78
Rollover	106	8	88	37.56
Other run-off-road	70	22	49	6.16
Pedestrian/bicycle/animal	56	37	-6	0.06
Other single vehicle	6	9	-142	2.78
Culpable multi-vehicle	108	44	34	3.64
Other multi-vehicle***	69	40		
All non-control group involvements	430	166	38	8.96
All crash involvements			28	
Total	583	261		
Single-vehicle crashes**	197	45	63	23.32

* For the 2x2 table formed by this row and the "control group" row.

** excludes pedestrian, bicycle, animal crashes.

*** includes involvements that are not culpable but are not part of the control-group.

Effectiveness confidence bounds in various fatal crash involvements

Now that we have discussed ESC effectiveness in various fatal crash involvements in passenger cars and LTVs, let us discuss further the confidence bounds for certain effectiveness estimates that are statistically significant. The entries in Table 11 will be used to produce an approximate sense of sampling errors and confidence bounds. As previously explained, the two numbers in the table, a and b, are counts of non-relevant crash involvements (control group) while the other two numbers, c and d, represent counts of relevant involvements in various crash modes – in vehicles that are equipped with ESC versus earlier versions of similar make-models that are not. These numbers can be considered independent Poisson variates.[13] As discussed, the effectiveness statistic in various crashes

$$E = 1 - \hat{\theta} = 1 - \left[\left(\frac{d}{c} \right) \div \left(\frac{b}{a} \right) \right]$$

is based sample odds ratio, $\hat{\theta}$.

The sampling distribution of the odds ratio can be highly skewed even for moderately large sample sizes. On the other hand, the log transform of the sample odds ratio, $\log \hat{\theta}$,

---

[13] Agresti, A. (2002) *Categorical Data Analysis*, Wiley, New York, pp.16-25.

22

has a less skewed sampling distribution and thus, is more symmetric. For large sample sizes, the distribution can be approximate to a normal distribution with a mean of log θ and a standard deviation often referred to as asymptotic standard error and denoted by ASE, of

$$ASE(\log\hat{\theta}) = \sqrt{\left[\left(\frac{1}{a}\right)+\left(\frac{1}{b}\right)+\left(\frac{1}{c}\right)+\left(\frac{1}{d}\right)\right]}$$

Taking 1.645 standard deviations on either side of the log $\hat{\theta}$ yields approximate confidence bounds (two-sided $\alpha = .10$, i.e., 90 percent bounds):

$$\log\hat{\theta} \pm 1.645\,ASE(\log\hat{\theta})$$

The confidence intervals can then be transformed back by using the exponential function to form the confidence interval for the odds ratio and subtracting from 1 to calculate effectiveness estimates.

Tables 14 and 15 show not only the estimates and chi-square statistics but also the confidence bounds of the effects of ESC for the following crash involvements: all run-off-road, rollover, culpable multi-vehicle, single-vehicle (excluding pedestrian, bicycle, animal crashes), and all crash involvements – in passenger cars and LTVs, respectively.

In Table 14, the effectiveness statistic in all run-off-road crashes is $1 - 0.6438 = 0.3562$, which is based on a sample odds ratio $(154/217) / (183/166) = 0.6438$. The natural log of the sample odds ratio equals $\log(0.64) = -0.4404$ and its asymptotic standard error equals $[1/166 + 1/183 + 1/217 + 1/154]^{.5} = 0.1503$. A 90 percent confidence interval for the sample log odds ratio equals $-0.4404 \pm 1.645(0.1503)$, or $(-0.6877, -0.1932)$. The corresponding confidence interval for the sample odds ratio is

$$\left[\exp(-0.6877), \exp(-0.1932)\right] = (0.5027, 0.8243).$$

Consequently, the confidence bounds for the effectiveness estimate is

$$\left[(1-0.8243)*100, (1-0.5027)*100\right] = (18 \text{ percent}, 50 \text{ percent}).$$

Using the same approach, 69 percent effectiveness in rollovers yields approximate confidence bounds (46 to 83 percent fatality reduction). A non-significant 36 percent increase in crashes involving pedestrians, bicycles, or animals suggests that the effectiveness of ESC is between -95 and 5 percent. The reduction estimates are between -4 and 37 percent for all culpable involvements in fatal multi-vehicle crashes. For all fatal crash involvements in passenger cars, the confidence bounds for the reduction (14 percent) are estimated to be 0 to 26 percent. For a 36 percent reduction in single-vehicle crashes involving passenger cars (excluding pedestrian, bicycle, animal crashes), the confidence bounds are between 18 and 50 percent.

23

Due to higher chi-square statistics for the effectiveness estimates (smaller standard deviations) in certain fatal crash involvements in LTVs (specifically in all run-off-roads and rollovers), the confidence bounds are tighter in LTVs than in passenger cars. For instance, in all fatal run-off-roads, the confidence bounds for the fatality reduction are 56 to 80 percent in LTVs as compared to 18 to 50 percent in passenger cars. Similarly, the effect of ESC in reducing fatal rollover crashes is so large that the confidence bounds range from 77 to 94 percent. Since the observed effect of ESC in reducing crashes involving pedestrians, bicycles, or animals crashes is negligible and non-significant, the confidence interval is between -60 and 29 percent. The reduction in culpable fatal multi-vehicle crash involvements in LTVs, on the other hand, is not as large as the reductions in various run-off-road involvements; thus, the confidence bounds are between 6 and 54 percent. For all types of fatal crashes in LTVs, the 28 percent effectiveness yields approximate confidence bounds, 13 to 41 percent fatality reduction. Last but not least, the reduction in single-vehicle crashes (not including pedestrian, bicycle, animal crashes) for LTVs is expected to be somewhere between 48 and 74 percent.

Table 14: Effectiveness Confidence Bounds in Certain Fatal Crash Involvements in Passenger Cars

Type of fatal crash involvement	Vehicles with no ESC	Vehicles with ESC	Effectiveness	Confidence Bounds	Chi-square Value
	Fatal Crashes	Fatal Crashes	(%)	(%)	
Control group	*166*	*183*			
All run-off-road	217	154	36	18 to 50	8.62
Rollover	36	12	70	46 to 83	12.71
Pedestrian/bicycle/animal	46	69	-36	-95 to 5	1.99
Culpable multi-vehicle	176	157	19	-4 to 37	1.91
All crash involvements			14	0 to 26	
Single vehicle crashes*	223	157	36	18 to 50	9.04

* excludes pedestrian, bicycle, and animal crashes.

Table 15: *Effectiveness Confidence Bounds*
in Certain Fatal Crash Involvements in LTVs

Type of fatal crash involvement	Vehicles with no ESC	Vehicles with ESC	Effectiveness	Confidence Bounds	Chi-square Value
	Fatal Crashes	Fatal Crashes	(%)	(%)	
Control group	*153*	*95*			
All run-off-road	191	36	70	56 to 79	29.90
Rollover	106	8	88	77 to 94	37.56
Pedestrian/bicycle/animal	56	37	-6	-60 to 29	0
Culpable multi-vehicle	108	44	34	6 to 54	3.64
All crash involvements			28	13 to 41	
Single vehicle crashes*	197	45	63	48 to 74	23.32

* excludes pedestrian, bicycle, and animal crashes.

Thus far, we have seen that ESC is highly effective in reducing fatal crashes – not only in passenger cars but also in LTVs. Let us now look at the results from the State analyses to see if these effectiveness rates still hold for the State data analyzed. The State data samples include fatal and non-fatal crash involvements although the vast majority of the involvements are non-fatal (i.e., property damage, possible injury, non-incapacitating and incapacitating injury). With the State data, separate analyses were performed in each of the selected States. Let us start with the analysis results in California.

Crashes in California

(Passenger cars)
As previously explained, the California data sample only includes Mercedes-Benz vehicles, because we could not identify other vehicles due to lack of VIN information. As shown in Table 16, run-off-road crashes in California were reduced by 66 percent in passenger cars equipped with ESC, and the reduction (relative to the control group) is statistically significant with a chi-square value of 125.22. Rollover crashes alone were significantly decreased (76 percent effectiveness; chi-square statistic 15.54). Side impacts with fixed objects were included in "other run-off-road" involvements because we could not identify those crashes based on the available information coded in the data files. Crashes that involved pedestrians, bicycles, or animals in California were decreased by 73 percent, and the decrease is statistically significant. The reduction (22 percent) of culpable involvements in multi-vehicle crashes is also statistically significant with chi-square statistic 19.32. For all non-control group involvements, the decrease is 32 percent, and it is statistically significant (a chi-square value of 61.99). When considering all crash involvements in passenger cars in California, the result showed that the benefit is quite strong for ESC – 15 percent.

Table 16: *Effect of ESC on Crashes in Mercedes-Benz Passenger Cars*
(California: 2001-2003)

Type of crash involvement	Vehicles with no ESC	Vehicles with ESC	Effectiveness	Chi-square Value
	All Crashes	All Crashes	(%)	
Control group	*1688*	*2510*		
All run-off-road	331	167	66	125.22
Rollover	25	9	76	15.54
Other run-off-road	306	158	65	113.08
Pedestrian/bicycle/animal	10	4	73	5.65
Other single vehicle	48	60	16	0.78
Culpable multi-vehicle	874	1017	22	19.32
Other multi-vehicle*	151	183		
All non-control group involvements	1414	1431	32	61.99
All crash involvements			15	
Total	3102	3941		

*includes involvements that are not culpable but are not part of the control-group.

(LTVs – mostly SUVs)

As shown in Table 17, ESC is even more effective (81 percent) in reducing run-off-road crashes in LTVs than in passenger cars, and the reduction is statistically significant. Particularly, LTVs equipped with ESC decreased the risk of rolling over by 78 percent. Likewise, those vehicles also reduced the risk of hitting a pedestrian, bicycle, or an animal by 26 percent although the effect is not statistically significant. Culpable involvements of LTVs in multi-vehicle crashes in California were reduced by nearly one-fourth (24 percent). The reduction was even more significant in all culpable involvements as evidenced by the 40 percent effectiveness and 9.66 chi-square value. In general, ESC is equally as effective in reducing all crash involvements (16 percent) in LTVs as in passenger cars.

Table 17: _Effect of ESC on Crashes in Mercedes-Benz LTVs_
(California: 2001-2003)

Type of crash involvement	Vehicles with no ESC	Vehicles with ESC	Effectiveness	Chi-square Value
	All Crashes	All Crashes	(%)	
Control group	_106_	_575_		
All run-off-road	20	21	81	29.62
Rollover	5	6	78	7.18
Other run-off-road	15	15	82	24.13
Pedestrian/bicycle/animal	1	4	26	0.07
Other single vehicle	6	15	54	2.57
Culpable multi-vehicle	57	235	24	2.29
Other multi-vehicle*	8	48		
All non-control group involvements	92	323	35	7.60
All crash involvements			16	
Total	198	898		

*includes involvements that are not culpable but are not part of the control-group.

Crashes in Florida

(Passenger cars)

All run-off-road crashes in Florida were decreased by 28 percent for passenger cars –
Table 18. The risk of a rollover or a side impact with fixed object was also lowered in
vehicles with a factory installed ESC. In fact, the relative decreases were 14 percent and
43 percent for side impacts with fixed objects and rollovers, respectively. Furthermore,
the effects are statistically significant for all run-off-roads and rollovers. Collisions with
pedestrians, bicycles, or animals and culpable involvements in multi-vehicle crashes also
benefited from ESC as indicated in Table 18 with statistically significant 49 percent and
13 percent effectiveness, respectively. All culpable crashes decreased by 13 percent, and
the decrease is statistically significant. On the whole, ESC is effective in reducing all
crash involvements by 5 percent.

(LTVs – mostly SUVs)

Table 19 shows that ESC reduced all run-off-road crashes in LTVs by 66 percent with
chi-square statistic 50.08. The reduction is equally substantial in rollovers (78 percent
effectiveness and a chi-square value of 16.83). Crashes that involved a pedestrian,
bicycle, or an animal were also decreased (20 percent effectiveness), but the relative
effect is not statistically significant. On the other hand, the effect of ESC in reducing all
culpable involvements in multi-vehicle crashes is statistically significant as evidenced by
the 10 percent effectiveness and a chi-square value of 3.44. Similarly, ESC is quite

effective in reducing all non-control group crashes (a 13 percent reduction and chi-square statistic of 8.74). The overall effect of ESC in reducing all crash involvements in Florida is slightly higher for LTVs – 6 percent reduction.

Table 18: *Effect of ESC on Crashes in Passenger Cars*
(Florida: 1997-2003)

Type of crash involvement	Vehicles with no ESC	Vehicles with ESC	Effectiveness	Chi-square Value
	All Crashes	All Crashes	(%)	
Control group	*5104*	*5441*		
All run-off-road	422	324	28	18.59
Side impact with fixed object	63	58	14	0.64
Rollover	33	20	43	4.06
Other run-off-road	326	246	29	16.03
Pedestrian/bicycle/animal	82	45	49	13.13
Other single vehicle	401	471	-10	1.88
Culpable multi-vehicle	2259	2105	13	13.96
Other multi-vehicle*	906	856		
All non-control group involvements	4070	3801	12	19.71
All crash involvements			5	
Total	9174	9242		

* includes involvements that are not culpable but are not part of the control-group.

Table 19: *Effect of ESC on Crashes in LTVs*
(Florida: 1997-2003)

Type of crash involvement	Vehicles with no ESC	Vehicles with ESC	Effectiveness	Chi-square Value
	All Crashes	All Crashes	(%)	
Control group	*2993*	*1655*		
All run-off-road	257	48	66	50.08
Side impact with fixed object	36	13	35	1.74
Rollover	57	7	78	16.83
Other run-off-road	164	28	69	35.93
Pedestrian/bicycle/animal	34	15	20	0.53
Other single vehicle	298	135	18	3.41
Culpable multi-vehicle	1227	609	10	3.44
Other multi-vehicle*	517	314		
All non-control group involvements	2333	1121	13	8.74
All crash involvements			6	
Total	5326	2776		

* includes involvements that are not culpable but are not part of the control-group.

Crashes in Illinois

(Passenger cars)

Statistically significant reductions in all run-off-road crashes (53 percent), especially in rollovers (80 percent) were also found in Illinois for passenger cars equipped with ESC – Table 20. The results also showed significant declines not only in pedestrian, bicycle, and animal crashes (26 percent) but also in crash involvements where the vehicle was considered the culpable party in a multi-vehicle crash (16 percent). All non-control group involvements were reduced (13 percent) as well. When we looked at all crash involvements in Illinois for passenger cars, we found that ESC, by and large, is still fairly effective – 7 percent reduction.

Table 20: **Effect of ESC on Crashes in Passenger Cars**
(Illinois: 1997-2002)

Type of crash involvement	Vehicles with no ESC	Vehicles with ESC	Effectiveness	Chi-square Value
	All Crashes	All Crashes	(%)	
Control group	*3548*	*2636*		
All run-off-road	420	146	53	60.63
Rollover	27	4	80	11.15
Other run-off-road	393	142	51	25.53
Pedestrian/bicycle/animal	247	136	26	7.49
Other single vehicle	662	528	-7	1.24
Culpable multi-vehicle	2002	1253	16	17.57
Other multi-vehicle*	1244	899		
All non-control group involvements	4575	2962	13	15.56
All crash involvements			7	
Total	8123	5598		

* includes involvements that are not culpable but are not part of the control-group.

(LTVs – mostly SUVs)

Although the reduction in rollovers in Illinois – which is credited to ESC – was only a few percentage points higher (87 percent versus 80 percent) in LTVs than in passenger cars, the decrease in all run-off-road crashes was substantially higher in LTVs than in passenger cars (80 percent versus 53 percent, respectively) – Tables 20-21. The analysis results also revealed potential increase (-20 percent) in pedestrian, bicycle, and animal crashes, but such increase (if any) is not statistically significant since the chi-square statistic (0.88) does not exceed 3.84. Culpable involvements in multi-vehicle crashes were also improved (due to ESC) by 28 percentage points, and the improvement is statistically significant. All non-control group crash involvements were reduced by 22 percent as well, and the reduction is also statistically significant – Table 21. Once again, ESC is quite successful in reducing all crash involvements (13 percent) for LTVs.

Table 21: Effect of ESC on Crashes in LTVs
(Illinois: 1997-2002)

Type of crash involvement	Vehicles with no ESC	Vehicles with ESC	Effectiveness	Chi-square Value
	All Crashes	All Crashes	(%)	
Control group	*1252*	*959*		
All run-off-road	147	22	80	59.71
Rollover	41	4	87	21.44
Other run-off-road	106	18	78	40.18
Pedestrian/bicycle/animal	60	55	-20	0.88
Other single vehicle	217	165	1	0.00
Culpable multi-vehicle	756	416	28	19.71
Other multi-vehicle*	484	332		
All non-control group involvements	1664	990	22	18.52
All crash involvements			13	
Total	2916	1949		

* includes involvements that are not culpable but are not part of the control-group.

Crashes in Kentucky

(Passenger Cars)

As for Kentucky, Table 22 shows a significant 47 percent reduction in all run-off-road crash involvements of passenger cars equipped with ESC. Of that, the decrease in rollovers is 73 percent, although the effectiveness is statistically not significant (1.72 chi-square statistic) due to limited number of involvements in our data sample (i.e., 7 crashes total; 6 pre-ESC crashes and 1 post-ESC crash). The effectiveness in passenger car involvements with pedestrians, bicycles, or animals in Kentucky is negative (-4 percent) and non-significant (chi-square statistic of 0.02). As for culpable involvements of passenger cars in multi-vehicle crashes in Kentucky, the effect is positive but non-significant as well. Likewise, the effect of ESC on non-control group involvements is positive (10 percent). As a whole, ESC still has a positive influence on all crash involvements in Kentucky for passenger cars.

Table 22: _Effect of ESC on Crashes in Passenger Cars_
(Kentucky: 1997-2002)

Type of crash involvement	Vehicles with no ESC	Vehicles with ESC	Effectiveness	Chi-square Value
	All Crashes	All Crashes	(%)	
Control group	*771*	*482*		
All run-off-road	117	39	47	10.80
Rollover	6	1	73	1.72
Other run-off-road	111	38	45	9.59
Pedestrian/bicycle/animal	37	24	-4	0.02
Other single vehicle	30	19	-1	0.00
Culpable multi-vehicle	409	231	10	1.02
Other multi-vehicle*	165	112		
All non-control group involvements	758	425	10	1.68
All crash involvements			5	
Total	1529	907		

* includes involvements that are not culpable but are not part of the control-group.

(LTVs – mostly SUVs)

As for LTVs, ESC is highly effective in reducing all-run-off road crashes, especially rollovers – Table 23. The reductions are statistically significant with the effectiveness equals to 77 percent for all run-off-roads and 92 percent for rollovers. On the other hand, crashes that involved pedestrians, bicycles, or animals, were increased by 21 percent, but the increase is once again statistically not significant. Although the effectiveness in culpable involvements in multi-vehicle crashes is also not significant, it is nevertheless a strong positive effect (17 percent). The effect of ESC on all non-control group crashes is also positive (17 percent). All in all, ESC is very effective in reducing all LTV crash involvements in Kentucky as shown by the 9 percent reduction.

Table 23: Effect of ESC on Crashes in LTVs
(Kentucky: 1997-2002)

Type of crash involvement	Vehicles with no ESC	Vehicles with ESC	Effectiveness	Chi-square value
	All crashes	All crashes	(%)	
Control group	*463*	*208*		
All run-off-road	95	10	77	20.73
Rollover	28	1	92	10.08
Other run-off-road	67	9	70	12.15
Pedestrian/bicycle/animal	22	12	-21	0.28
Other single vehicle	16	6	17	0.14
Culpable multi-vehicle	257	96	17	1.60
Other multi-vehicle*	106	62		
All non-control group involvements	496	186	17	2.27
All crash involvements			9	
Total	959	394		

* includes involvements that are not culpable but are not part of the control-group.

Crashes in Missouri

(Passenger Cars)

All run-off-road crash involvements in Missouri were also decreased (44 percent) for passenger cars equipped with ESC – Table 24. Similarly, crash involvements with pedestrians, bicycles, or animals as well as culpable involvements in multi-vehicle crashes and all non-control group crashes were reduced by 48 percent, 8 percent, and 17 percent, respectively. Except for the reduction in culpable involvements in multi-vehicle crashes, all other reductions are statistically significant. We found that ESC was 10 percent effective in reducing all passenger car crash involvements.

(LTVs – mostly in SUVs)

A continuing trend in crash reductions – specifically in all run-off-road crashes – is observed for LTVs in Missouri – Table 25. The reduction is significant with 80 percent effectiveness and chi-square statistic of 39.51. ESC also reduced crashes that involved pedestrians, bicycles, or animals, but the reduction is statistically non-significant because the chi-square statistic is close to 0. The results also suggested that ESC provides little benefits (if any) to LTVs – particularly, in culpable involvements in multi-vehicle crashes, but for all non-control group crash involvement, the benefits are significant (a 16 percent reduction). Overall, all crash involvements decreased by 9 percent.

Table 24: Effect of ESC on Crashes in Passenger Cars
(Missouri: 1997-2003)

Type of crash involvement	Vehicles with no ESC	Vehicles with ESC	Effectiveness	Chi-square Value
	All Crashes	All Crashes	(%)	
Control group	*1141*	*915*		
All run-off-road	232	105	44	21.09
Side impact with fixed object	21	20	-19	0.30
Rollover	14	2	82	6.59
Other run-off-road	197	83	47	22.24
Pedestrian/bicycle/animal	58	24	48	7.43
Other single vehicle	161	98	24	4.15
Culpable multi-vehicle	757	560	8	1.28
Other multi-vehicle*	313	225		
All non-control group involvements	1521	1012	17	9.65
All crash involvements			10	
Total	2662	1927		

* includes involvements that are not culpable but are not part of the control-group.

Table 25: Effect of ESC on Crashes in LTVs
(Missouri: 1997-2003)

Type of crash involvement	Vehicles with no ESC	Vehicles with ESC	Effectiveness	Chi-square Value
	All Crashes	All Crashes	(%)	
Control group	*562*	*283*		
All run-off-road	150	15	80	39.51
Side impact with fixed object	13	0	100	6.50
Rollover	31	0	100	13.34
Other run-off-road	106	15	72	22.08
Pedestrian/bicycle/animal	27	12	12	0.12
Other single vehicle	71	36	-1	0.00
Culpable multi-vehicle	401	204	-1	0.01
Other multi-vehicle*	158	76		
All non-control group involvements	807	343	16	3.04
All crash involvements			9	
Total	1369	626		

* includes involvements that are not culpable but are not part of the control-group.

Crashes in Pennsylvania

(Passenger cars)

For passenger cars in Pennsylvania, ESC has quite an impact on all run-off-road involvements (40 percent effectiveness, 21.69 chi-square statistic), especially in rollovers (66 percent effectiveness, 5.23) – as shown in Table 26. Although it is not statistically significant, observed ESC effectiveness in pedestrian, bicycle, and animal crashes is negative – as illustrated in Table 26 by the negative 23 percent effectiveness. Pennsylvania is the first State whose analysis shows a negative ESC effectiveness in culpable involvements in multi-vehicle crashes, but the effect is statistically non-significant. Furthermore, ESC has little effect in reducing all crash involvements (1 percent) in Pennsylvania for passenger cars.

Table 26: **Effect of ESC on Crashes in Passenger Cars**
(Pennsylvania: 1997-2001, 2003)

Type of crash involvement	Vehicles with no ESC	Vehicles with ESC	Effectiveness	Chi-square Value
	All Crashes	All Crashes	(%)	
Control group	*983*	*553*		
All run-off-road	431	146	40	21.69
Side impact with fixed object	53	18	40	3.36
Rollover	26	5	66	5.23
Other run-off-road	352	123	38	16.61
Pedestrian/bicycle/animal	62	43	-23	1.04
Other single vehicle	31	24	-38	1.34
Culpable multi-vehicle	731	454	-10	1.53
Other multi-vehicle*	282	183		
All non-control group involvements	1537	850	2	0.06
All crash involvements			1	
Total	2520	1403		

* includes involvements that are not culpable but are not part of the control-group.

(LTVs – mostly SUVs)

Table 27 shows that ESC significantly reduced all run-off-roads in Pennsylvania by 63 percent for LTVs and was even more effective in reducing rollovers – as evidenced by the 86 percent effectiveness and 18.26 chi-square statistic. We see possible harm with ESC for crash involvements with pedestrians, bicycles, or animals; however, the effect is not statistically significant. This analysis showed that ESC has negligible effect on culpable involvements of LTVs in multi-vehicle crashes in Pennsylvania, but it has quite an effect on all non-control group involvements (22 percent) as well as on crash involvements (14 percent). These effects are statistically significant.

Table 27: *Effect of ESC on Crashes in LTVs*
(Pennsylvania: 1997-2001, 2003)

Type of crash involvement	Vehicles with no ESC	Vehicles with ESC	Effectiveness	Chi-square Value
	All Crashes	All Crashes	(%)	
Control group	*355*	*210*		
All run-off-road	193	42	63	28.64
Side impact with fixed object	19	8	29	0.63
Rollover	48	4	86	18.26
Other run-off-road	126	30	60	17.71
Pedestrian/bicycle/animal	19	19	-69	2.49
Other single vehicle	13	3	61	2.27
Culpable multi-vehicle	226	132	1	0.01
Other multi-vehicle*	129	70		
All non-control group involvements	580	266	22	4.97
All crash involvements			14	
Total	935	476		

* includes involvements that are not culpable but are not part of the control-group.

<u>Crashes in Wisconsin</u>

(Passenger cars)

Similar to all run-off-road involvements of passenger cars in other States, the involvements in Wisconsin were significantly decreased – 34 percent effectiveness and 9.58 chi-square statistic – Table 28. In addition to involvements with pedestrians, bicycles, or animals, culpable involvements in multi-vehicle crashes as well as in all non-control group crashes decreased – for passenger cars equipped with ESC versus those not equipped with ESC – as shown by the 23 percent, 16 percent, and 17 percent, respectively. The decreases are significant in the culpable involvements in multi-vehicle crashes and in all non-control group crashes. ESC is effective overall in Wisconsin for passenger cars – as demonstrated in Table 28 with an 11 percent reduction of all crash involvements.

(LTVs – mostly SUVs)

Likewise, LTVs with ESC are much less at risk of being involved in run-off-road crashes when compared to vehicles with no ESC – as shown in Table 29 by the 68 percent

35

effectiveness and 32.45 chi-square value. In addition to these single-vehicle crash involvements, other culpable involvements in multi-vehicle crashes were also significantly decreased by 27 percent – though the decrease is not as substantial as in the single-vehicle crashes. ESC had negligible effect on crashes that involved pedestrians, bicycles, or animals as evidenced by the negative 4 percent effectiveness and 0.02 chi-square statistic. All in all, ESC is very effective in reducing all crash involvements in Wisconsin – 23 percent for LTVs.

Table 28: *Effect of ESC on Crashes in Passenger Cars*
(Wisconsin: 1997-2003)

Type of crash involvement	Vehicles with no ESC	Vehicles with ESC	Effectiveness	Chi-square Value
	All Crashes	All Crashes	(%)	
Control group	*705*	*516*		
All run-off-road	214	104	34	9.58
Rollover	13	4	58	2.42
Other run-off-road	201	100	32	8.19
Pedestrian/bicycle/animal	120	68	23	2.49
Other single vehicle	57	25	40	4.39
Culpable multi-vehicle	645	395	16	4.28
Other multi-vehicle*	178	147		
All non-control group involvements	1214	739	17	6.14
All crash involvements			11	
Total	1919	1255		

* includes involvements that are not culpable but are not part of the control-group.

Table 29: *Effect of ESC on Crashes in LTVs*
(Wisconsin: 1997-2003)

Type of crash involvement	Vehicles with no ESC	Vehicles with ESC	Effectiveness	Chi-square Value
	All Crashes	All Crashes	(%)	
Control group	*359*	*231*		
All run-off-road	162	33	68	32.45
Rollover	50	4	88	21.51
Other run-off-road	112	29	60	17.15
Pedestrian/bicycle/animal	42	28	-4	0.02
Other single vehicle	26	10	40	1.85
Culpable multi-vehicle	310	146	27	5.68
Other multi-vehicle*	128	63		
All non-control group involvements	668	280	35	15.16
All crash involvements			23	
Total	1027	511		

* includes involvements that are not culpable but are not part of the control-group.

Crashes in all selected States

At this point, we have comprehensively discussed separate analysis results of the effect of ESC in reducing relevant crash involvements of passenger cars and LTVs – relative to the control-group involvements – in each selected State. Let us now try to assess the overall effect of ESC across these seven States. The easiest way to obtain a single estimate would be to pool the cases from the States and perform a single analysis. We are reluctant to analyze the data that way because:
Different States have different crash-reporting thresholds. States with low thresholds have more reported crash cases per capita, and would account for an unjustifiably large share of the pooled data.
The make-model mix varies from State to State.
The distribution of crash types varies from State to State, as do the definitions of the crash types.

Instead, we will use the _weighted_ mean of the sample log odds ratios from the seven States as the best indicator of the central tendency of the data, and the _weighted_ standard error of the log odds ratios as a basis for judging statistical significance and estimating confidence bounds, which will be discussed in the next section. In this section, we will concentrate only on the overall effects of the systems.[14]

The reason for using the mean of the log odds ratios is due to the facts that it is (1) less affected by extreme values, (2) useful as a measure of central tendency for certain positively skewed distributions, and (3) an appropriate measure to use for averaging rates (e.g., crash reduction/increase). The weighted mean is used to take into account the difference in the sampling distributions among the seven States. The weighted value for each State is simply the reciprocal of the sum of the reciprocals of the four frequencies from the 2x2 contingency table. The weighted mean of the log odds ratios is determined using the PROC MEANS procedure in SAS from the seven log odds ratios and the weighted value for each of the seven States.

(Passenger cars)

We will first discuss the effects on passenger cars and then on LTVs. Table 30 shows not only the individual percentage reductions in the ratios of relevant involvements to control-group involvements – in passenger cars with ESC versus cars of similar make-models without ESC – in each State, but also the overall reductions – which are determined by computing the weighted mean of the sample log odds ratios. For example, for all run-off-road crashes in California involving passenger cars (Table 16), the odds ratio equals $[(167/331) \div (2510/1688)] = 0.3393$. The log odds ratio is $\log (0.3393) = -1.0809$. The weighted value equals

$$1 \div \left\{ \left(\frac{1}{1688} \right) + \left(\frac{1}{2510} \right) + \left(\frac{1}{331} \right) + \left(\frac{1}{167} \right) \right\} = 100.00$$

---

[14] Charles Farmer recommended this approach in his peer review.

37

Using the same approach, the log odds ratios for Florida, Illinois, Kentucky, Missouri, Pennsylvania, and Wisconsin, are -0.3282, -0.7595, -0.6289, -0.5720, -0.5073, and -0.4095, respectively. The corresponding weighted values are 171.36, 101.10, 26.62, 63.28, 83.37, and 56.67. The computed weighted mean of the log odds ratios from SAS program is -0.5969. The weighted mean of the overall effectiveness is estimated simply by first taking the antilogarithm of -0.5969 and then subtracting [exp (-0.5969) = 0.5505] from 1, which yields 45 percent effectiveness.

As illustrated in Table 30 in the gray-highlighted column, ESC is most effective in reducing all run-off-road crashes by 45 percent, particularly in rollovers by 64 percent. Other single-vehicle crash involvements – where ESC also had an impact – are those that involved pedestrians, bicycles, or animals, even though the results are inconsistent across States. Nevertheless, the impact is still positive: 26 percent effectiveness. The table shows negative overall effectiveness (-2 percent) of ESC in reducing "other single-vehicle" crashes (such as collisions with parked cars, impacts with thrown or falling objects, undercarriage scrapes, or first-event fires), and the results are also inconsistent among the States. Culpable involvements in multi-vehicle crashes received some benefit from ESC as evidenced by the 13 percent effectiveness. When all crashes (including culpable, non-culpable, and the control-group involvements) were taken into account, ESC is still quite effective – as shown by the 8 percent effectiveness. In the State analyses, we also analyzed the effectiveness of ESC on single-vehicle crashes that did not involve pedestrians, bicycles, or animals. As shown in Table 30, the reductions of single-vehicle crashes were observed in each of the States, and the overall reduction (which is the weighted mean of the reductions in seven States) is 26 percent.

Table 30: The Effects and Mean of the Effects of ESC on Various Crash Involvements of Passenger Cars – By State

Type of crash involvement	CA	FL	IL	KY	MO	PA	WI	*Mean of States
	Effectiveness (%)	Effectiveness (%)	Effectiveness (%)	Effectiveness (%)	Effectiveness (%)	Effectiveness (%)	Effectiveness (%)	Effectiveness (%)
Control group								
All run-off-road	66	28	53	47	44	40	34	45
Rollover	76	43	80	73	82	66	58	64
Other run-off-road	65	29	51	45	47	38	32	45
Pedestrian/bicycle/animal	73	49	26	-4	48	-23	23	26
Other single vehicle**	16	-10	-7	-1	24	-38	40	-2
Culpable multi-vehicle	22	13	16	10	8	-10	16	13
All crash involvements	15	5	7	5	10	1	11	8
Single vehicle crashes (excluding pedestrian, bike, animal crashes)	60	9	16	37	36	35	35	26

* indicates the weighted mean of the effects in all 7 States.

** includes collisions with parked vehicles.

39

(Light-Trucks – mostly SUVs)

ESC was even more effective in reducing run-off-road crashes (relative to the control-group involvements) for LTVs, especially in rollover involvements – as shown in Table 31 by the 72 percent and 85 percent effectiveness, respectively. In calculating the weighted mean effect for the rollover reductions in seven States, the crash involvements in Illinois and neighboring Missouri were pooled because a 100 percent reduction in Missouri yields an undefined log odds ratio. ESC had a negative effect on pedestrian, bicycle, or animal crashes. In LTVs, culpable involvements in other single- and multi-vehicle crashes were also decreased by 12 percent and 16 percent, respectively. Overall, ESC was also effective in reducing all crash involvements in LTVs. The reduction in single-vehicle crashes (excluding those involving pedestrians, bicycles, or animals) involving LTVs is not as high as the reduction in all run-off-road crashes (48 percent versus 72 percent.

Table 31: The Effects and Mean of the Effects of ESC on Various Crash Involvements in LTVs - By State

Type of crash involvement	CA	FL	IL	KY	MO	PA	WI	*Mean of States
	Effectiveness (%)	Effectiveness (%)	Effectiveness (%)	Effectiveness (%)	Effectiveness (%)	Effectiveness (%)	Effectiveness (%)	Effectiveness (%)
Control group								
All run-off-road	81	66	80	77	80	63	68	72
Rollover**	78	78	87	92	100	86	88	85
Other run-off-road	82	69	78	70	72	60	60	69
Pedestrian/bicycle/animal	26	20	-20	-21	12	-69	-4	-11
Other single vehicle	54	18	1	17	-1	61	40	12
Culpable multi-vehicle	24	10	28	17	-1	1	27	16
All crash involvements	16	6	13	9	9	14	23	10
Single vehicle crashes (excluding pedestrian, bike, animal crashes)	74	40	33	68	54	63	64	48

* indicates the weighted mean of the effects in all 7 States.

** Data in IL and MO were pooled to allow computation of a log-odds ratio.

41

Effectiveness confidence bounds and statistical significance in various crash involvements

Now, let us discuss confidence intervals for certain effectiveness estimates that are considered significant crash reductions in passenger cars and LTVs – Tables 32 and 33. In the State analysis results, we will calculate confidence bounds using weighted mean of the log odds ratios, weighted standard errors based on the variation of the seven States, and take 1.943 standard errors for two-sided $\alpha = 0.10$, i.e., 90 percent bounds, because 1.943 is the 95th percentile of the t-distribution with (7-1)=6 degrees of freedom. A general notation for calculating the confidence interval for the overall effect is illustrated below by the formula[15]:

$$\text{Confidence interval} = (\text{overall effect})_{weighted\,mean\,of\,\log\,odds} \pm t_{05,N-1} (SE)$$

where

1. $(\text{overall effect})_{weighted\,mean\,of\,\log\,odds}$ is determined using the approach discussed in the previous section.
2. $t_{05,N-1}$ is the t value for a two-sided test (90 percent bounds) and N-1 degrees of freedom.
3. N is the number of individual effects.
4. SE is the weighted standard error of the log odds ratios obtained from the PROC MEANS procedure in SAS (i.e., the weighted standard deviation of these ratios divided by $\sqrt{7}$, the square root of the number of States).

Let us take one example (all run-off-road crash involvements of passenger cars) and discuss the statistics in detail. As discussed in the previous section, the weighted mean of the log odds ratios is -0.5969. The weighted standard error of the log odds ratios (computed from SAS) is 0.1068. Applying these statistics to the above t-confidence interval equation produces

90% confidence interval = -0.5969 $\pm$ 1.943 (0.1068) = -08044 to -0.3894

which translates to (0.4474, 0.6775) in odds ratio by taking the antilogarithm of the confidence interval estimates (-0.8044, -0.3894). The confidence interval for the overall effect of ESC on all run-off-road crashes involving passenger cars is (32%, 55%). Using the same approach, we found the following confidence intervals for other critical reductions in passenger cars: 50 to 75 percent in rollover crashes, 8 to 41 percent in crashes that involved a pedestrian, bicycle, or an animal, 7 to 18 percent in culpable involvements in multi-vehicle crashes, and 5 to 10 percent in all crashes. For single-vehicle crash involvements (not including collisions with a pedestrian, bicycle, or an animal), the expected effectiveness is between 10 and 40 percent. The overall

---

[15] *Handbook of Probability and Statistics with Tables*, Second Edition, McGraw-Hill, Inc., USA, 1970, pp. 244-245.

effectiveness estimates listed in Table 32 are statistically significant because the confidence bounds are greater than 0.

For LTVs, the confidence bounds for the reductions (listed in Table 33) were also calculated using the above method. Table 33 shows that the weighted mean of the effectiveness of ESC in reducing all LTV run-off-crashes is 72 percent, and the 90 percent confidence interval is 65 to 77 percent. As previously discussed, because the crash involvements in Illinois and neighboring Missouri were pooled for the analysis of rollovers (Table 31), the number of individual effects *(N)* is reduced to 6, and the critical *t*-value for 6-1 = 5 degrees of freedom is 2.015[16]. Hence, the confidence interval for the rollover reduction is (79%, 90%). A negative 11 percent effectiveness in crashes involving pedestrians, bicycles, or animals implies that the expected effectiveness estimate is between -31 and 6 percent. The confidence bounds for the decrease in culpable involvements in multi-vehicle crashes are 7-23 percent. The confidence bounds for the reduction in all crashes in LTVs are between 6 and 14 percent. When we look at all single-vehicle crashes except the pedestrian, bicycle, and animal crashes, the confidence interval for the reduction is (35%, 58%). In general, the spread of the sample distribution determines the range of the confidence interval. In other words, if the effectiveness estimates are consistent across all studied States, then the spread of the sample distribution is expected to be small, and hence, the range of the confidence interval is expected to be small as well – as evidenced by the estimates and the confidence bounds listed in Tables 32-33. Similar to the effects on passenger cars, the weighted mean effects on LTVs for run-off-road, rollover, culpable multi-vehicle, and all crash involvements are statistically significant because the confidence bounds are all positive. The overall reduction of single-vehicle crashes excluding pedestrian, bicycle, and animal crashes is also statistically significant. Only the overall effect on crashes that involved a pedestrian, bicycle, or an animal is not statistically significant because the estimates from the individual States are inconsistent.

---

[16] *Handbook of Probability and Statistics with Tables*, Second Edition, McGraw-Hill, Inc., USA, 1970, p. 383.

Table 32: *The Effects and Mean of the Effects of ESC of Various Crash Involvements of Passenger Cars - By State*

Type of crash involvement	CA	FL	IL	KY	MO	PA	WI	*Mean of States	Confidence bounds
	Effectiveness (%)	Effectiveness (%)	Effectiveness (%)	Effectiveness (%)	Effectiveness (%)	Effectiveness (%)	Effectiveness (%)	Effectiveness (%)	Effectiveness (%)
All run-off-road	66	28	53	47	44	40	34	45	32 to 55
Rollover	76	43	80	73	82	66	58	64	50 to 75
Pedestrian/ bike/animal	73	49	26	-4	48	-23	23	26	8 to 41
Culpable multi-vehicle	22	13	16	10	8	-10	16	13	7 to 18
All crash involvements	15	5	7	5	10	1	11	8	5 to 10
Single-vehicle crashes**	60	9	16	37	36	35	35	26	10 to 40

* indicates the weighted mean of the effectiveness estimates of 7 States.

** excludes pedestrian, bicycle, animal crashes.

Notes:
1) Bold – weighted mean effect is statistically significant.
2) Plain type – weighted mean effect is not statistically significant.

44

Table 33: The Effects and Mean of the Effects of ESC on Various Crash Involvements in LTVs – By State

Type of crash involvement	CA	FL	IL	KY	MO	PA	WI	*Mean of States	Confidence bounds
	Effectiveness (%)	Effectiveness (%)	Effectiveness (%)	Effectiveness (%)	Effectiveness (%)	Effectiveness (%)	Effectiveness (%)	Effectiveness (%)	Effectiveness (%)
All run-off-road	81	66	80	77	80	63	68	**72**	65 to 77
Rollover**	78	78	87	92	100	86	88	**85**	79 to 90
Pedestrian/bike/animal	26	20	-20	-21	12	-69	-4	-11	-31 to 6
Culpable multi-vehicle	24	10	28	17	-1	1	27	**16**	7 to 23
All crash involvements	16	6	13	9	9	14	23	**10**	6 to 14
Single-vehicle crashes***	74	40	33	68	54	63	64	**48**	35 to 58

* indicates the weighted mean of the effectiveness estimates of 7 States.

** A combined effectiveness calculated from crash involvements in IL and MO as 1 State (Tables 21 and 25). E = {1-[((4+0) / (959+283)) ÷ ((41+31) / (1252+562))]}=0.92.

Only 6 States (*N=6*) are included for this calculation because the data in IL and MO data were pooled.

*** excludes pedestrian, bicycle, animal crashes.

Notes:

1) Bold – weighted mean effect is statistically significant.

2) Plain type – weighted mean effect is not statistically significant.

45

LOGISTIC REGRESSION ANALYSIS

Logistic regression permits a refinement and check of the basic analyses. Logistic regression by the GENMOD procedure[17] was used to estimate the effect of passenger car ESC on the probability that a crash involvement was relevant (specifically the involvement of a vehicle running off the road) as opposed to being non-relevant (i.e., the control-group involvement), while controlling for other factors. Estimating the impact of ESC in reducing single-vehicle run-off-road crashes could be confounded by factors related to the driver, vehicle, roadway, or other circumstances or by a proportionately different make-model mix in the vehicles without ESC and the vehicles with ESC. To make certain that the effect of ESC is estimated accurately, variables were included in the logistic regression to control for those external factors, other than ESC, that could possibly influence the ratio of run-off-road to control-group crash involvements.

Here is one example of how demographic factors (i.e., driver characteristics) could confound the effect of ESC. Since ESC is more commonly installed on higher-priced vehicles, ESC-equipped vehicles are more likely to be driven by a certain group of drivers – perhaps more affluent and older – than by other segments of the driving population. Although it is unlikely that the majority of the passenger cars used in our analysis sample were driven by 16-24 year-olds, we included this group of drivers in our model because they could potentially confound the effect of ESC in reducing run-off-road crashes. This driver-age category generally has a higher percentage of drivers considered more aggressive in their driving than others, and thus, their vehicles are at a higher risk of being involved in run-off-road crashes. Furthermore, based on historical trends of drivers involved in crashes by sex and crash severity,[18] it is a common stereotype and many researchers have attempted to test the hypothesis that – male drivers are more likely to be involved in crashes because they are generally more aggressive in their driving than female drivers. Thus, driver gender was included in the regression model as well.

Other external factors include whether or not the crash occurred: (1) at night, (2) on a rural roadway, or (3) on a freeway. These factors could influence the proportion of run-off-road crash involvements because a driver, in those environment or situations, is more likely to lose control of a vehicle, and hence, ESC is more likely to be activated. The vehicle make-model group was also included in the regression model. The age of the vehicle was initially considered, but we did not include this factor in the final regression analysis, because its effect was not statistically significant when (1) all independent variables mentioned above were included in the regression model and even when (2) only variables for ESC and vehicle age were included.[19]

The data points in the logistic regressions for passenger cars and LTVs are the FARS driver fatality cases, each given a weight factor of 1. Cases with unreported age, gender, or roadway function class (rural or urban) are excluded; so are the drivers younger than 14 years or older than 97 years. The dependent variable ROR (for run-off-road) – a dichotomous variable with a

---

[17] Allison, P.D. (1999) *Logistic Regression Using the SAS System: Theory and Application*, Cary, NC: SAS Institute Inc., pp.82-83.

[18] *Traffic Safety Facts,* 2004, NHTSA Technical Report No. DOT HS 809 919, Washington D.C., 2004, pp.19-20.

[19] Charles Farmer, in his peer review, recommended analysis (2) as an additional way to check if vehicle age is a factor. Claes Tingvall also emphasized the potential importance of vehicle age in the current analysis, and especially in future analyses as the vehicles without ESC get older.

binomial distribution – equals 1 if the crash involvement was a single-vehicle run-off-road and 2 if it was non-relevant (control group). The key independent variable, ESCS indicates whether or not the crash-involved vehicle had ESC as standard equipment: 1 if ESC was installed and 0 if it was not. The other independent variables are:

Gender, expressed as DRVMALE (= 1 if the driver is male and 0 if the driver is female).

The driver's age, expressed as D16_24 (= 1 if the driver is 16-24 years old and 0 otherwise).

Time of day, expressed as NITE (= 1 if 7:00 p.m. – 5:59 a.m., 0 if 6:00 a.m. – 6:59 p.m.)

Roadway function class (RURAL, FREEWAY). RURAL has the value 1 if the crash occurred on a rural roadway, 0 if the crash occurred on an urban roadway. FREEWAY has the value 1 if the crash occurred on an interstate, freeway, or expressway and 0 otherwise.

The vehicle make-model group, expressed as a categorical variable (CAR_MFG).

The regression for the analysis of passenger cars is based on 708 FARS fatality cases. The likelihood-ratio (LR) statistics for TYPE 3 analysis[20] are shown in Table 34. TYPE 3 analysis was used because of the inclusion of the CLASS variable CAR_MFG in the model. After we controlled for the important quantifiable human parameters – driver's age, gender – plus other variables that are strongly associated with single-vehicle run-off-road crash involvements, the effect of ESC on fatal run-off-road involvements (relative to the control-group involvements) in passenger cars is statistically significant with the LR chi-square statistic being 18.39.[21] In fact, the model predicted that ESC reduced fatal single-vehicle run-off-road crashes by $1 - \exp(-0.7838) = 54$ percent because the regression coefficient estimate for ESCS is -0.7838. The results from the 2x2 contingency table analysis (Table 12) showed a 36 percent reduction in all fatal run-off-road crash involvements relative to the same control-group involvements used in the regression analysis. Thus, if we compare those two results, we see that the 2x2 contingency table approach may be more conservative than the regression analysis method. Hence, the regression analysis gives us increased confidence that the effectiveness estimates obtained from our 2x2 contingency table analyses are not biased in favor of ESC.

The regression results from the passenger car analysis supported our assumptions that: (1) male drivers are at a higher risk of being involved in fatal run-off-road crashes than females, and (2) the 16-24 year-old group of drivers is also at a higher risk of being involved in fatal run-off-road crashes when compared to other groups of drivers. These factors are statistically significant – as indicated by the LR statistics shown in Table 34 (chi-square values 35.45 for DRVMALE and 19.65 for D16_24). Other independent variables (specifically those related to the environment) that also had statistically significant effects on fatal run-off-road crashes are those that indicate whether or not the crash occurred at night (NITE) or on a rural roadway (RURAL) – as evidenced by the LR chi-square statistics 46.13 and 19.23, respectively. The effect of another factor FREEWAY – which indicates whether or not the crash occurred on a freeway, interstate, or expressway – is borderline-significant (chi-square = 6.46; Pr>chisq = 0.0111). The vehicle make-model group (CAR_MFG) is, however, statistically significant – as shown by the 23.14 LR chi-square statistic and 0.0008 Pr>chisq.

---

[20] Allison, P.D. (1999) *Logistic Regression Using the SAS System: Theory and Application*, Cary, NC: SAS Institute Inc., p. 27.
[21] Results are statistically significant at the 0.05 level when the LR chi-square statistic is greater than 3.84.

Table 34: _Likelihood-ratio Statistics for TYPE 3 Analysis of the Independent Variables (FARS: 1997-2004) - Passenger Cars_

Source	Degrees of Freedom	Chi-Square	Pr > ChiSq
ESCS	1	18.39	<.0001
DRVMALE	1	35.45	<.0001
D16_24	1	19.65	<.0001
NITE	1	46.13	<.0001
RURAL	1	19.23	<.0001
FREEWAY	1	6.46	0.0111
CAR_MFG	6	23.14	0.0008

For the analysis of LTVs, the regression is based on 471 FARS fatality cases. In this analysis, the model predicted that ESC reduced fatal single-vehicle run-off-road crashes involving LTVs by 68 percent because the regression coefficient estimate for ESCS is -1.1320. Similar to the effects of ESC and other independent variables in the passenger car regression analysis, the effects of those variables in the LTV analysis (shown in Table 35) are also statistically significant except for the effect of the vehicle make-model group. Again, the effectiveness estimate obtained from the regression analysis is similar to the observed estimate from the 2x2 contingency table analysis – 68 percent versus 70 percent, respectively (Table 13).

Table 35: _Likelihood-ratio Statistics for TYPE 3 Analysis of the Independent Variables (FARS: 1997-2004) - LTVs_

Source	Degrees of Freedom	Chi-Square	Pr > ChiSq
ESCS	1	21.31	<.0001
DRVMALE	1	11.89	0.0006
D16_24	1	10.75	0.0010
NITE	1	7.51	0.0061
RURAL	1	18.37	<.0001
FREEWAY	1	21.83	<.0001
CAR_MFG	2	1.19	0.5511

COMPARISON OF 2-CHANNEL AND 4-CHANNEL ESC SYSTEMS

We also compared crash reduction with 2-channel versus 4-channel ESC systems for reasons already explained in the "background" section. The 4-channel ESC systems have a control algorithm and means to apply all four brakes <u>individually</u> whereas the 2-channel systems do not. In other words, the 2-channel systems are capable of applying brake torque only to the two front wheels. Hence, the latter systems only have the capability to correct oversteer whereas the former systems are capable of correcting not only oversteer but also understeer.

Table 35 shows passenger car fatalities (single vehicle run-off-road versus control-group involvements) in vehicles with and without ESC – by system type (2-channel versus 4-channel systems). We subdivided our FARS data into two groups of make-models: those equipped with 2-channel ESC systems (all GM models selected for the study except the Corvette) and those equipped with 4-channel ESC systems (all non-GM models selected for the study plus the Corvette). The make-models used in this analysis (2- versus 4-channel ESC) were identical to those listed in Tables 2 through 7. There were no make-models that had received both types. The crash reduction with 2-channel ESC is compared to cars of the same make-models, but in the model years before they received ESC – and likewise to the make-models that received 4-channel ESC.

In the 2-channel systems (Category 1), there were 56 fatalities in single-vehicle run-off-road crashes and 56 fatalities in non-relevant crashes (control group) – in vehicles without ESC, a risk ratio of 1.000. In vehicles with ESC, there were 16 run-off-road and 41 non-relevant fatalities, a risk ratio of 0.390. That is a significant 61 percent reduction in single-vehicle run-off-road fatalities in 2-channel systems relative to the control-group. In the 4-channel systems (Category 2), single-vehicle run-off-road fatalities also decreased, but the decrease is 39 percent and statistically significant. Additional analyses were conducted to refine and properly characterized these results, as discussed below.

We can test if the fatality reduction is significantly greater in category 1 (2-channel) than in category 2 (4-channel) by performing a three-dimensional contingency table analysis. The difference in effectiveness is not statistically significant, as evidenced by the chi-square of the three-way interaction term when the CATMOD procedure of SAS is applied to the three-way table.[22] In fact, the chi-square statistic for the difference in the reduction in fatal single-vehicle run-off-road crashes between the two ESC systems is 1.35, which suggests that the larger observed fatality reduction with 2-channel systems (shown in Table 35) is not significantly different from the observed reduction with the 4-channel system. The reductions in fatal single-vehicle run-off-road fatalities should be interpreted with caution because of the small number of fatalities in both samples (2-channel and 4-channel systems) – Table 35. As we have found from the FARS and State data analyses (discussed in detail in the previous sections), the single-vehicle run-off-road fatality reduction (due to ESC) for passenger cars is 36 percent (Table 12), and the run-off-road crash reduction (mostly non-fatal crashes) is 45 percent (Table 30). Based on these results, both fatal and non-fatal single-vehicle run-off-road reductions are expected to be somewhere in this range – whether the reduction is associated with 2-channel or 4-channel systems. Hence, a 61 percent reduction in fatal single-vehicle run-off-road involvements in vehicles with 2-channel systems is probably an unrealistic estimate and most certainly influenced by the small sample. Consequently, the difference in the reduction in fatal single-vehicle run-off-road crashes between the two ESC systems is found to be statistically non-significant (chi-square value of 1.35) – due to the small samples.

---

[22] For a CATMOD analysis of crash data with statistically significant three-way terms see Morgan, C., *The Effectiveness of Retroreflective Tape on Heavy Trailers*, NHTSA Technical Report No. DOT HS 809 222, Washington, 2001, pp. 29-37, summarized in Kahane (2004), pp. 43-44; *SAS/STAT® User's Guide*, Vol. 1, Version 6, 4th Ed., SAS Institute, Cary, NC, 1990.

Given that the N of FARS cases is insufficient for statistically meaningful results, we strongly believe, in this instance, that findings based on State data are more relevant and should be given greater weight.

Table 35: _Effectiveness of ESC as the Percentage Reduction in Fatal Single Vehicle Run-Off-Road Crash Involvements Relative to the Control-Group Involvements - by ESC System Type - for Passenger Cars_

	All run-off-road involvements	Control-group involvements	Risk Ratio	ESC reduction
Category 1: 2-channel ESC systems				
Vehicles not equipped with 2-channel	56	56	1.000	61%
Vehicles equipped with 2-channel	16	41	0.390	
Category 2: 4-channel ESC systems				
Vehicles not equipped with 4-channel	165	105	1.571	39%
Vehicles equipped with 4-channel	139	144	0.965	

Tables 36 shows not only the reduction in all single-vehicle run-off-road crashes – which consist of mostly non-fatal involvements – in each selected State, but also the weighted mean of the reductions in all States (similar to the calculation method used in Tables 30 and 31) – in passenger cars for 2-channel and 4-channel ESC systems. To test for the significance of the difference in effectiveness in six States, a four-dimensional CATMOD analysis was used. In theory, a four-dimensional CATMOD analysis involves aggregating data from all the States, using all possible combinations of the four variables in the model. However, this approach could create biased results due to: (1) different crash-reporting thresholds among the States, (2) variation of make-model mix from State to State, as well as (3) variation in the distribution of crash types from State to State. Thus, to adjust for those biases, a CATMOD analysis was performed on 6 x 4 x 2 table of State by ESC system by crash involvement. The dichotomous dependent variable is ROR with values (= 1 for single vehicle run-off-road involvements; = 2 for the control-group involvements). The independent variables are STATE (a categorical variable with six categories: 1, 2, 3, 4, 5, and 6 for six States – Florida, Illinois, Kentucky, Missouri, Pennsylvania, and Wisconsin, respectively), FOUR_CH (a dichotomous variable with values = 1 for 4-channel systems; = 2 for 2-channel systems), and ESC (another dichotomous variable with values = 1 for vehicles with ESC: 2-channel or 4-channel; = 2 for vehicles without ESC). The model included the following terms: 1) ROR*FOUR_CH, 2) ROR*ESC, 3) ROR*STATE, 4) ROR*FOUR_CH*STATE, 5) ROR*ESC*STATE, and 6) ROR*FOUR_CH*ESC.[23] The chi-square statistic for the ROR*FOUR_CH*ESC term was 4.69, which is statistically significant at the 0.05 level but not significant at the 0.025 level. In other words, the difference in

---

[23] The four-way interaction term is not included because we do not anticipate that the effectiveness gap between 4-channel and 2-channel systems would vary from State to State.

effectiveness between the 2-channel and 4-channel systems – specifically the larger observed reductions in single-vehicle run-off-road involvements with 4-channel systems versus the smaller observed reductions with 2-channel systems – are statistically significant in the State data.

Table 36: *Effectiveness of 2-Channel versus 4-Channel ESC Systems as the Percentage Reduction in All Single-Vehicle Run-Off-Road Crashes in Passenger Cars – By State Run-Off-Road Crash Reduction (%) in*

	Florida	Illinois	Missouri	Kentucky	Pennsylvania	Wisconsin	*Mean of the reductions in 6 States
With 2-channel ESC systems	0	49	31	37	50	11	30
With 4-channel ESC systems	31	57	48	50	41	47	43

* indicates the weighted mean of the reductions in 6 States.

Another approach that we used to compare 2-channel and 4-channel systems is to *pool* crash involvements (specifically single-vehicle run-off-roads and non-relevant involvements) from the States and estimate the significance of the difference in the overall effectiveness estimate of the two systems. Table 37 shows the sum of crash involvements from six States and the effectiveness estimates for the two systems. Again, the 4-channel systems are more effective in reducing single-vehicle run-off-road crashes than the 2-channel systems as shown by the 48 percent and 33 percent effectiveness, respectively. When the CATMOD procedure is applied to the three-way table, the chi-square of the three-way interaction term is 5.41, which is statistically significant not only at the 0.05 level but also at the 0.025 level. Regardless of how we assess the statistical significance of the difference in the effects of the two ESC systems in the State data analysis, the results from two different methods of analysis (non-parametric and parametric approach) consistently showed that the 4-channel systems reduced more single-vehicle crashes than the 2-channel systems.

Table 37: *Effectiveness of ESC as the Percentage Reduction in Single-Vehicle Run-Off-Road Crash Involvements Relative to the Control-Group Involvements from Six States - by ESC System Type - for Passenger Cars*

	All run-off-road involvements	Control-group involvements	Risk Ratio	ESC reduction
Category 1: 2-channel ESC systems				
Vehicles not equipped with 2-channel	437	3304	0.132	33%
Vehicles equipped with 2-channel	145	1638	0.089	
Category 2: 4-channel ESC systems				
Vehicles not equipped with 4-channel	1399	8948	0.156	48%
Vehicles equipped with 4-channel	719	8905	0.081	

CONCLUSIONS

Based on the analysis results from the 2x2 contingency tables, ESC appears to be extremely successful in reducing not only fatal crashes but also other crash involvements. Tables 38 and 39 summarize the effectiveness of ESC – as the percentage reduction in the ratios of relevant crash involvements to non-relevant involvements (i.e., the control-group) – in passenger cars and LTVs with ESC versus those without ESC, for certain crash involvements that are considered relevant.

In Table 38, ESC reduced all fatal single-vehicle run-off-road crash involvements by 36 percent in passenger cars and 70 percent in LTVs. The table also includes effectiveness estimates of fatal single-vehicle crashes excluding pedestrian, bicycle, animal crashes for passenger cars and LTVs – 36 percent and 63 percent, respectively. These results are similar to those in the 2004 study – which showed that ESC was effective in reducing fatal single-vehicle crashes by 30 percent in passenger cars and 63 percent in SUVs. The control group used in the 2004 study included all multi-vehicle crash involvements, whereas this study used non-relevant involvements – where the crash-involved vehicle was stopped, parked, traveled at a speed of less than 10 mph, or non-culpable party in a multi-vehicle crash on a dry road – as the control group. As mentioned in the 2004 study, using multi-vehicle crashes as the control group, when it is possible that multi-vehicle crashes are being reduced by ESC, actually means that the true effectiveness of ESC could be higher than what we had estimated in the 2004 study for fatal single-vehicle crashes. Based on the results from both studies, we may conclude with confidence that the effectiveness still hold for all fatal single-vehicle crashes especially for all run-off-road involvements in passenger cars and LTVs. Furthermore, the logistic regression analysis results showed that ESC reduced fatal run-off-road involvements by 54 percent for passenger cars and 68 percent for LTVs, controlling for certain external factors that could confound the effects of ESC. Initially, the vehicle "age" effect was included in the regression analysis, but the effect was not statistically significant even when only independent variables for ESC and vehicle age

52

were considered and other factors were not. Thus, this factor was not included in the final regression model. While the regression results showed that the age of vehicles had no significant effect on crash involvements (specifically fatal run-off-road involvements), it is important that this factor be evaluated in future analyses as ESC equipped vehicles get older and its influence on crash involvements might become more significant.

Moreover, in this study, we found that rollover involvements in fatal crashes decreased by 70 percent in passenger cars and 88 percent in LTVs. Hence, ESC is very effective in reducing fatal single-vehicle run-off-road crashes and extremely beneficial in rollover involvements. The reductions are statistically significant. Furthermore, multi-vehicle crash involvements were also reduced – in passenger cars and in LTVs with ESC – as shown by the positive effectiveness of 19 percent (although not statistically significant) and 34 percent, respectively – Table 38. The observed effects of ESC on pedestrian, bicycle, and animal crashes are not statistically significant, and they are inconsistent – as evidenced by the negative 36 percent and negative 6 percent effectiveness for passenger cars and LTVs, respectively. If we excluded pedestrian, bicycles, and animal crashes, ESC is very successful in reducing single-vehicle crashes – 36 percent for passenger cars and 63 percent for LTVs. The results are similar to those found in the 2004 study. Even with the inconsistent results for ESC in crashes that involved pedestrians, bicycles, or animals, ESC is still highly effective in reducing all fatal crash involvements (including the control-group involvements as well as other non-relevant involvements) by 14 percent in passenger cars and 28 percent in LTVs.

Table 38: *Effectiveness of ESC as the Percentage Reduction in Certain Fatal Relevant Crash Involvements in Passenger Cars and LTVs*

Types of Crash Involvement	Passenger Cars	LTVs
Single-vehicle crashes*	**36%**	**63%**
All run-off-road	**36%**	**70%**
Rollover	**70%**	**88%**
Pedestrian/bicycle/animal	-36%	-6%
Culpable multi-vehicle	19%	**34%**
All crashes	**14%**	**28%**

* *excludes pedestrian, bicycle, animal crashes.*
Notes:
1) Bold – effect is statistically significant.
2) Plain type – effect is not statistically significant.

Table 39 shows the weighted mean of the effects of ESC in all selected States (California, Florida, Illinois, Kentucky, Missouri, Pennsylvania, and Wisconsin) on certain relevant crash involvements (which consist of mostly non-fatal crashes) in passenger cars and LTVs. The notations listed in the table below are similar to those in Table 38. Similar to the results in the fatal crash analysis, ESC is highly effective in reducing all single-vehicle run-off-road crashes, especially rollovers. In fact, the reductions are 45 percent in passenger cars and 72 percent in LTVs for all run-off-road crashes – and 64 percent and 85 percent, respectively, for rollovers. Also, the observed reductions in single-vehicle crashes (not including pedestrian, bicycle, animal crashes) for passenger cars and LTVs are 26 percent and 48 percent, respectively, which are comparable (to some extent) to the results from the 2004 study (35 percent for passenger cars and 67 percent for SUVs). The results for ESC-equipped vehicles in crashes that involved a pedestrian, bicycle, or animal are inconsistent – as shown by the 26 percent and negative 11 percent effectiveness in passenger cars and LTVs, respectively. This contrasts with the fatal crashes, where the results were less favorable for cars. Culpable multi-vehicle crash involvements decreased by 13 percent in passenger cars and 16 percent in LTVs, and the decreases are statistically significant. When we looked at all crash involvements, we found significant reductions with ESC – 8 percent reduction in passenger cars and 10 percent reduction in LTVs.

Table 39: *Effectiveness of ESC as the Percentage Reduction in Certain Relevant Crash Involvemetns in Passenger Cars and LTVs*

Types of Crash Involvement	*Passenger Cars	*LTVs
Single-vehicle crashes**	**26%**	**48%**
All run-off-road	**45%**	**72%**
Rollover	**64%**	**85%**
Pedestrian/bicycle/animal	26%	-11%
Culpable multi-vehicle	**13%**	**16%**
All crashes	**8%**	**10%**

* indicates the weighted mean of the effectiveness estimates of 7 States.
** excludes pedestrian, bicycle, animal crashes.
Notes:
1) Bold – mean effect is statistically significant.
2) Plain type – mean effect is not statistically significant.

All in all, ESC significantly reduced single-vehicle crashes especially run-off-road and rollover involvements. In all likelihood, ESC may also be helpful in reducing culpable involvements in multi-vehicle crashes. At this time we do not have enough data for statistically meaningful results on the effect (if any) of ESC on pedestrian, bicycle, and animal crashes. Thus, we will continue to monitor the effect on those crashes in the future.

Although there is a fairly large variation in ESC effectiveness (possibly because the estimates somewhat depend on the choice of the control group) between different studies (as previously discussed in the "Background" Section), ESC is still highly effective in all studies. ESC is likely to have the largest effect on crashes involving severe injuries. Minor crashes are less likely to be rollovers. Thus, including a large number of damage only crashes or slight injury crashes in this study would likely reduce the effectiveness of ESC. In other words, the effectiveness of ESC may even be higher than estimated if the large number of damage-only crashes are excluded in the analysis. Although the sample used in this study is based on mostly luxury vehicles, we believe that ESC would still be highly effective across the entire on-road fleet based on the results from other studies (specifically those in Europe) where a large population of various vehicle classes were used in analyzing the effectiveness of ESC.

REFERENCES

Aga, M. and Okada, A. (2003) Analysis of Vehicle Stability Control (VSC)'s Effectiveness from Accident Data, Paper Number 541, *Proceedings of the 18th International Technical Conference on the Enhanced Safety of Vehicles*.

Agresti, A. (2002) *Categorical Data Analysis*, Wiley, New York, pp.16-25.

Allison, P.D. (1999) Logistic Regression Using the SAS System: Theory and Application, Cary, NC: SAS Institute Inc., pp.82-83.

Dang, J. (2004) *Preliminary Results Analyzing the Effectiveness of Electronic Stability Control (ESC) Systems*, NHTSA Evaluation Note No. DOT HS 809 790, Washington, DC.

Farmer, C. (2004) Effect of Electronic Stability Control on Automobile Crash Risk, *Traffic Injury Prevention*, Vol 5, pp. 317-325.

Green, P. and Woodrooffe, J. (2006) *The Effectiveness of Electronic Stability Control on Motor Vehicle Crash Prevention*, Report Number UMTRI-2006-12, University of Michigan Transportation Research Institute, Ann Arbor, MI.

Handbook of Probability and Statistics with Tables, Second Edition, McGraw-Hill, Inc., USA, 1970, pp. 244-245.

Lie, A., Tingvall, C., Krafft, M., and Kullgren, A. (2006) The Effectiveness of ESC (Electronic Stability Control) in Reducing Real Life Crashes and Injuries, *Traffic Injury Prevention*, Vol 7, pp. 38-43.

Lie, A., Tingvall, C., Krafft, M., and Kullgren, A. (2004) The Effectiveness of ESP (Electronic Stability Program) in Reducing Real Life Accidents, *Traffic Injury Prevention,* Vol 5, pp. 37-41.

Morgan, C. (2001) *The Effectiveness of Retroreflective Tape on Heavy Trailers*, NHTSA Technical Report No. DOT HS 809 222, Washington, DC, pp. 29-37.

SAS/STAT® User's Guide, Vol. 1, Version 6, 4th Ed., SAS Institute, Cary, NC, 1990.

Traffic Safety Facts, 2004, NHTSA Technical Report No. DOT HS 809 919, Washington, DC, pp.19-20.